T0135498

Event-based state-feedback control

Vom Promotionsausschuss der
Fakultät für Elektrotechnik und Informationstechnik
an der Ruhr-Universität Bochum
zur Erlangung des Grades
eines Doktor-Ingenieurs
genehmigte Dissertation

von Daniel Lehmann
aus Essen

Bochum, 2011

1. Gutachter: Prof. Dr.-Ing. Jan Lunze

2. Gutachter: Prof. Dr. Lars Grüne

Eingereicht am: 11. Mai 2011
Tag der mündlichen Prüfung: 22. Juli 2011

Bibliografische Information der Deutschen Nationalbibliothek

Die Deutsche Nationalbibliothek verzeichnet diese Publikation in der Deutschen Nationalbibliografie; detaillierte bibliografische Daten sind im Internet über http://dnb.d-nb.de abrufbar.

©Copyright Logos Verlag Berlin GmbH 2011
Alle Rechte vorbehalten.

ISBN 978-3-8325-2940-6

Logos Verlag Berlin GmbH
Comeniushof, Gubener Str. 47,
10243 Berlin
Tel.: +49 (0)30 42 85 10 90
Fax: +49 (0)30 42 85 10 92
INTERNET: http://www.logos-verlag.de

Acknowledgements

This thesis is the result of four years of research at the Institute of Automation and Computer Control (ATP) of Prof. Dr.-Ing. Jan Lunze at the Ruhr-Universität Bochum, Germany. The work would have not been possible without the support and encouragement of several persons whom I would like to thank in the following.

First of all, I would like to express my special gratitude to Prof. Dr.-Ing. Jan Lunze for giving me the chance to do this work at his institute and for his excellent supervision, support and inspirations. He had and still has a strong influence on my professional and personal development.

I also thank Prof. Dr. Lars Grüne who gave me a lot of valuable insights in control theory from a more mathematical perspective which definitely has helped a lot, and Prof. Dr.-Ing. Andreas Steimel, Prof. Dr.-Ing. Jürgen Oehm and Prof. Dr. Jörg Schwenk for accepting to take part in my examination.

Moreover, I am very grateful to my colleagues in Bochum: Prof. Dr.-Ing. Christian Schmid, Dr.-Ing. Johannes Dastych, Andrej, Axel, Christian, Jan, Jan, Jörg, Leila, Michael, Michael, Ozan, Piotr, Plinio, Rene, Sebastian, Thorsten and Yannick for a stimulating working atmosphere, proofreading my thesis and all the joint activities apart from working.

A special thanks goes to Andrea Marschall for helping me with the production of the figures in this work and in several other publications.

Besides the working environment, my deepest thankfulness goes to my parents Doro and Michael, my sister Inga and my brother Philip for perpetually motivating and supporting me.

Finally, there is one person remaining who was absolutely indispensable for me in this time, my wife Steffi. Thank you so much for your unconditional patience, permanent encouragement and just for always being there.

Bochum, August 2011 Daniel Lehmann

Contents

Abstract

Event-based control has been recently investigated as a means to reduce the communication over the feedback link in networked control systems. This thesis deals with a state-feedback approach to event-based control which allows approximating a continuous-time state-feedback loop with arbitrary precision while adapting the communication over the feedback link to the effect of exogenous disturbances.

The focus of this thesis lies in complementing the event-based state-feedback control by deriving new properties of the event-based state-feedback loop, proposing alternative methods for the analysis and improving the existing components by incorporating a dynamical controller and an augmented disturbance estimation. Moreover, the influences of uncertainties about the plant and imperfect communication links on the behaviour of the event-based control loop are investigated and suitable strategies are proposed to overcome these circumstances.

It is shown that, in all these cases, a stable behaviour of the event-based state-feedback loop can be retained and a bound on the approximation error with respect to the behaviour of the continuous-time state-feedback loop can be determined. However, the performance of the event-based control loop may deteriorate when considering non-ideal plant information and communication imperfections which, moreover, generally increase the communication in the loop.

The event-based state-feedback control is applied to a thermofluid process both in simulations and experiments. The experimental evaluation validates the theoretical results by showing that the concept is robust against severe uncertainties about the plant parameters and that event-based control is capable of significantly reducing the information exchange over the feedback link compared to conventional discrete-time control.

Deutsche Kurzfassung

Struktur und Zielstellung der ereignisbasierten Regelung

Der in dieser Arbeit betrachtete ereignisbasierte Regelkreis ist in Abb. 1 dargestellt. Er besteht aus

- der Regelstrecke mit der Eingangsgröße $\boldsymbol{u}(t)$, der Ausgangsgröße $\boldsymbol{y}(t)$, dem Zustand $\boldsymbol{x}(t)$, und der Störung $\boldsymbol{d}(t)$,
- dem Ereignisgenerator, der abhängig vom Systemverhalten die Ereigniszeitpunkte t_k $(k = 0, 1, ...)$ bestimmt, zu denen der Zustand $\boldsymbol{x}(t_k)$ an den Stellgrößengenerator übertragen wird, und
- dem Stellgrößengenerator, der unter Verwendung dieser Informationen das kontinuierliche Stellsignal $\boldsymbol{u}(t)$ berechnet.

Die durchgezogenen Linien in der Abbildung kennzeichnen einen kontinuierlichen Datenaustausch, wohingegen die gestrichelte Linie einen nur zu den Ereigniszeitpunkten t_k stattfinden Informationsaustausch anzeigt.

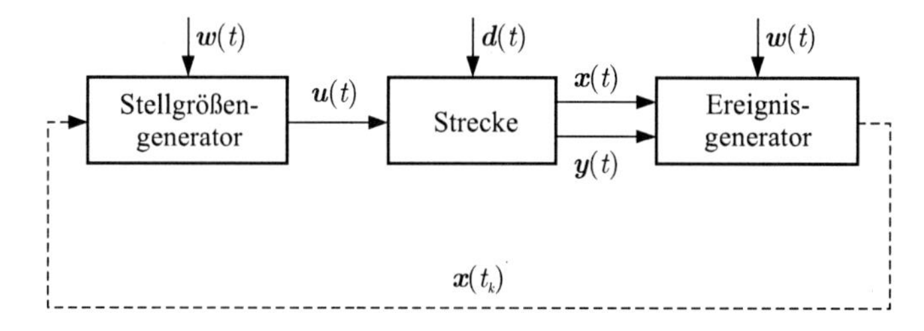

Abbildung 1.: Ereignisbasierter Regelkreis

Das Ziel ereignisbasierter Regelungen ist die Reduktion der Kommunikation über den Rückführzweig des Regelkreises, indem nur dann Daten übertragen werden, wenn ein Ereignis eine nicht tolerierbare Abweichung des tatsächlichen Regelkreisverhaltens vom gewünschten Verhalten signalisiert. Der Datenaustausch im Regelkreis wird dadurch auf Zeitintervalle beschränkt, in denen der Regler auf Störungen oder Änderungen von Sollwerten reagieren muss, um das spezifizierte Regelungsziel zu erfüllen.

Das Hauptanwendungsgebiet der ereignisbasierten Regelung liegt bei digital vernetzten dynamischen Systemen. Dabei soll die ereignisbasierte Regelung die Kommunikation der Regelkreiskomponenten über das Netzwerk reduzieren, um die bei Netzüberlastung entstehenden Verzögerungen und Paketausfälle zu verhindern. In dieser Arbeit werden darüber hinaus noch eine Vielzahl weiterer Anwendungsfelder aufgeführt, in denen die ereignisbasierte Regelung Vorteile bringt.

Die ereignisbasierte Regelung wirft neue theoretische Fragen auf, weil die Grundannahme der Abtastregelung, dass der Informationsaustausch zwischen Regelstrecke und Regler zeitperiodisch erfolgt, nicht erfüllt ist. Die wesentlichen Fragestellungen sind:

- Was ist ein geeignetes Regelungsziel für ereignisbasierte Regelungen?
- Was ist eine geeignete Struktur für den ereignisbasierten Regelkreis und wie sind die Komponenten des ereignisbasierten Regelkreises zu entwerfen?
- Welche Daten müssen zu den Ereigniszeitpunkten übertragen werden?

Die erste Fragestellung wird im Folgenden exemplarisch diskutiert.

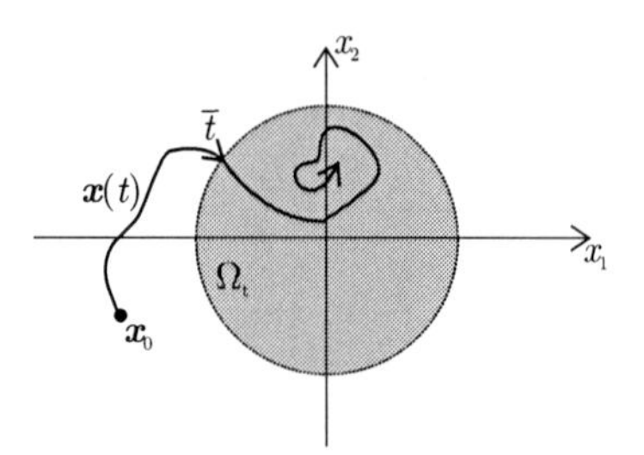

Abbildung 2.: Praktische Stabilität

Asymptotische Stabilität und praktische Stabilität. Da der Informationsaustausch im ereignisbasierten Regelkreis abhängig vom Systemverhalten nur dann erfolgt, wenn das Verhalten einen Toleranzbereich verlässt, kann Stabilität im Sinne der asymptotischen Stabilität durch ereignisbasierte Regelungen grundsätzlich nicht erreicht werden. Eine im Kontext der ereignisbasierten Regelung besser geeignete Stabilitätsdefinition liefert die so genannte praktische Stabilität („Ultimate Boundedness"). Ein System heißt ultimativ beschränkt, wenn jede Trajektorie ausgehend von einem Anfangszustand $\boldsymbol{x}_0$ in ein Zielgebiet Ω_t gebracht werden kann und dort trotz Störungen für alle zukünftigen Zeiten verbleibt (Abb. 2). Diese Stabilitätsdefinition bildet die Grundlage für alle Stabilitätsuntersuchungen im Rahmen dieser Arbeit.

Hauptergebnisse der Arbeit

Die in dieser Arbeit behandelte ereignisbasierte Zustandsrückführung hat das folgende Ziel:

> Die ereignisbasierte Zustandsrückführung soll in der Lage sein, eine kontinuierliche Zustandsrückführung mit einstellbarer Güte und einstellbarer Kommunikationshäufigkeit zu approximieren.

Der vorgestellte Ansatz unterscheidet sich durch drei signifikante Neuheiten von der bestehenden Literatur:

1. Der Stellgrößengenerator ist nicht mehr nur durch ein Halteglied 0-ter Ordnung beschrieben. Stattdessen wird zwischen zwei Ereigniszeitpunkten eine Steuerung in der offenen Wirkungskette verwendet, um das Eingangssignal $\boldsymbol{u}(t)$ kontinuierlich anzupassen. Dafür nutzt der Stellgrößengenerator ein Model der kontinuierlichen Zustandsrückführung.

2. Der Ereignisgenerator enthält eine Kopie des kontinuierlichen Regelkreismodells und generiert Ereignisse immer dann, wenn der Zustand der Regelstrecke im nicht mehr tolerierbarem Maße von dem Modellzustand abweicht.

3. Sowohl der Stellgrößengenerator als auch der Ereignisgenerator nutzen einen Störungsschätzer, um das Modellverhalten an das tatsächliche Verhalten anzupassen.

Durch den Entwurf der Komponenten des ereignisbasierten Regelkreises gemäß dieser Merkmale weist die ereignisbasierte Zustandsrückführung folgende Eigenschaften auf:

- Der Approximationsfehler zwischen dem Zustand der ereignisbasierten Zustandsrückführung und dem angestrebten Zustand der kontinuierlichen Zustandsrückführung ist beschränkt und kann beliebig eingestellt werden (Kapitel 3, Theorem 3).

- Die Kommunikationshäufigkeit über den Rückführzweig des ereignisbasierten Regelkreises ist beschränkt. Sie hängt von der Größe der Störung ab und kann durch die Ereignisschranke ebenfalls beliebig vorgeben werden. Bei der Wahl der Ereignisschranke muss allerdings ein Kompromiss zwischen einer gewünschten Approximationsgenauigkeit und einer akzeptablen Kommunikationshäufigkeit gefunden werden (Kapitel 3, Theorem 5).

Der Beitrag dieser Arbeit liegt in der Erweiterung des bestehenden ereignisbasierten Grundkonzepts. Erstens ergänzt die Arbeit das Konzept durch alternative Analysemethoden, neue Eigenschaften des ereignisbasierten Regelkreises und Verbesserungen der bestehenden Komponenten, d. h.

- eine alternative Stabilitätsanalyse (Abschnitt 3.4.2, Theorem 4),
- eine Untersuchung des Führungsverhaltens der ereignisbasierten Zustandsrückführung (Abschnitt 3.4.5, Theorem 6) und die Verbesserung des Führungsverhaltens durch die Verwendung eines dynamischen Reglers im Stellgrößengenerator (Abschnitt 4.3, Theorem 12),
- Verbesserungen der Störungsschätzung (Abschnitt 4.4),
- eine gemeinsame Grundlage für den Vergleich zwischen der ereignisbasierten Zustandsrückführung und der zeitdiskreten Zustandsrückführung (Abschnitt 3.5, Theorem 8), und
- eine zeitdiskrete Realisierung der Komponenten des ereignisbasierten Regelkreises (Abschnitt 4.5, Theoreme 13, 14).

Zweitens werden die Konsequenzen nicht-idealer Eigenschaften der Regelstrecke bzw. der Kommunikation untersucht und Möglichkeiten zur Kompensation dieser Effekte vorgestellt. Dies umfasst die Ausarbeitung von

- Strategien, die die Kompensation von Unsicherheiten in der Regelstrecke (Modellunsicherheiten, nicht messbare Zustandsvariablen) ermöglichen (Abschnitte 3.4.6, 4.2, 4.4, Theoreme 7, 9), und
- geeigneten Anpassungen der Komponenten des ereignisbasierten Regelkreises, um nichtideale Kommunikationseigenschaften wie Verzögerungen, Paketausfälle und Beschränkungen der Datenrate besser handhaben zu können (Kapitel 5, Theoreme 15, 16).

In all diesen Fällen wird gezeigt, dass der ereignisbasierte Regelkreis stabil bleibt und eine obere Schranke für den Approximationsfehler im Bezug auf die kontinuierliche Zustandsrückführung berechnet werden kann. Die Betrachtung nicht-idealer Effekte zeigt jedoch, dass sich die Güte der Regelung im Hinblick auf die garantierte Approximationsgenauigkeit verschlechtern kann. Erwartungsgemäß erhöhen diese Effekte die Kommunikationshäufigkeit, wobei weiterhin Grenzen für ein maximales Kommunikationsaufkommen existieren (Theoreme 10, 17, 18). Die Analyse in dieser Arbeit konzentriert sich dabei auf Regelkreise, bei denen angenommen wird, dass dem Aktorkonten und dem Sensorknoten alle Aktorsignale bzw. alle Messsignale zur Verfügung stehen.

Drittens wird im Rahmen dieser Arbeit ein thermofluider Prozess definiert und genutzt, um die ausgearbeiteten Konzepte in Simulationen und Experimenten zu erproben. Die experimentelle Erprobung validiert, dass

- die Konzepte robust gegenüber Modellunsicherheiten sind und
- nicht-ideale Effekte eher die Kommunikationshäufigkeit erhöhen als die Performanz des ereignisbasierten Regelkreises verschlechtern. Die Kommunikationshäufigkeit liegt aber allgemein deutlich unter dem Informationsaustausch einer zeitdiskreten Regelung.

1. Introduction

1.1. Event-based control

1.1.1. Structure

In theory and practice, there are two control paradigms which are used almost exclusively when considering the control of continuous-time systems: *continuous-time control* [70, 79, 118] and *discrete-time control* [32, 83, 96]. Their primary difference is given by the fact that in the continuous-time control loop the information between the sensors, the controller and the actuators is transmitted continuously, whereas in the discrete-time control loop a communication between these components is invoked periodically, i.e. at equidistant instances of time (sampling times).

This thesis presents a new control scheme which is called *event-based control* [8, 28, 84]. It can be characterised as follows:

> In contrast to the continuous-time or discrete-time control strategy, in event-based closed-loop systems, an information exchange among the sensors, the controller and the actuators is triggered in dependence upon the system behaviour, e.g. when the system variables exceed certain tolerance bounds. Hence, the sampling and the information flow in the feedback loop are adapted to the current needs and, therefore, generally occur not equidistantly in time.

In other words, the activity of the controller is restricted to time intervals in which the controller inevitably must act in a closed-loop manner in order to guarantee desired specifications of the closed-loop behaviour.

The structure of the event-based control loop is depicted in Fig. 1.1. It consists of

- the plant with input vector $\boldsymbol{u}(t)$, output vector $\boldsymbol{y}(t)$, state vector $\boldsymbol{x}(t)$, and disturbance vector $\boldsymbol{d}(t)$,
- the event generator which determines the event times t_k at which information $\boldsymbol{s}(t_k)$ is

sent towards the controller,

- the controller which computes the signal $\boldsymbol{u}(t_k)$ based on the information obtained at time t_k and the command input $\boldsymbol{w}(t)$, and
- the control input generator which accordingly determines the continuous-time input $\boldsymbol{u}(t)$ of the plant.

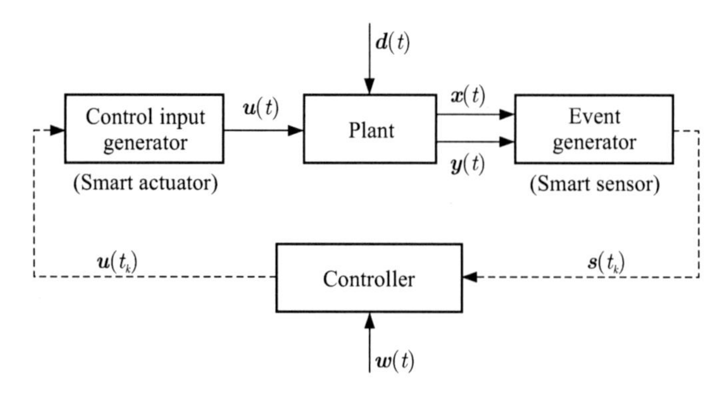

Figure 1.1.: Event-based control loop

In the figure, the solid lines indicate continuous-time signals, whereas the dashed lines indicate that these information links are only used at event times t_k $(k = 0, 1, ...)$. In order to realise the proposed structure, the control input generator and the event generator have to be implemented on smart actuators and smart sensors, respectively, which are components with built-in processing units [71]. Throughout this thesis, it is assumed that the processing power of these components does not impose any restrictions on the computational complexity of the control input generator and the event generator.

Note that the structure of event-based control distinguishes from the discrete-time control loop [83] by replacing the sampler by the event generator and the hold by the control input generator which allows a more involved specification of the control input $\boldsymbol{u}(t)$ between two consecutive events.

1.1.2. Application fields

The main reason for generally using continuous-time control or discrete-time control in practical applications is the well established theory for the analysis and the design of the control loop. However, these approaches also have some fundamental drawbacks:

- A continuous-time information transfer is only possible if the control loop is implemented by analogue hardware. As nowadays almost all controllers are implemented on digital hardware, the continuous-time control theory can only be applied if certain conditions in terms of sampling time and plant properties are satisfied [83].

- In the discrete-time control loop, the control is executed driven by a clock and, thus, independently of the system behaviour. Consequently, during time intervals in which the system variables do not change significantly and no information feedback is necessary to meet the performance requirements on the plant, e.g. in steady state, computation and communication resources are wasted unnecessarily [27].

- In the opposed situation in which unpredictable events, e.g. exogenous disturbances, affect the system, discrete-time control is not capable to immediately counteract these effects as it firstly has to wait for the next sampling time given by the external clock. In this case, an inadmissible degradation of the system performance can only be avoided by using a suitably short sampling interval which, however, might be again inappropriate in the sense of wasting computation and communication resources.

Besides these general aspects, there are a lot of application fields in which the consideration of event-based control is beneficial or even indispensable. Here, networked control systems can be seen as the main reason for the recent interest in event-based control which likewise holds for the work presented in this thesis.

Networked control systems. Control systems, in which the control loop is closed over a digital communication medium, are called *networked control systems* (NCS) [34, 62, 66, 123, 131, 132]. Compared to the traditional point-to-point architecture of a wired communication, a digital network offers several benefits with respect to lower costs, a simplified installation and maintenance. As a central feature, it additionally allows an almost unlimited flexibility in setting up and changing the required communication infrastructure. Hence, the communication links between the relevant nodes of the network can be simply adapted to the current needs.

The general structure of NCS is depicted in Fig. 1.2 consisting of several (sub-)systems, which may be physically interconnected, and the respective sensor (S), actuator (A) and controller (C) nodes. All these nodes may be widespread within the entire system and can be connected in an arbitrary way through the digital communication medium. Again, the solid lines indicate continuous-time signals but the dashed lines now indicate that these information links can be used flexibly.

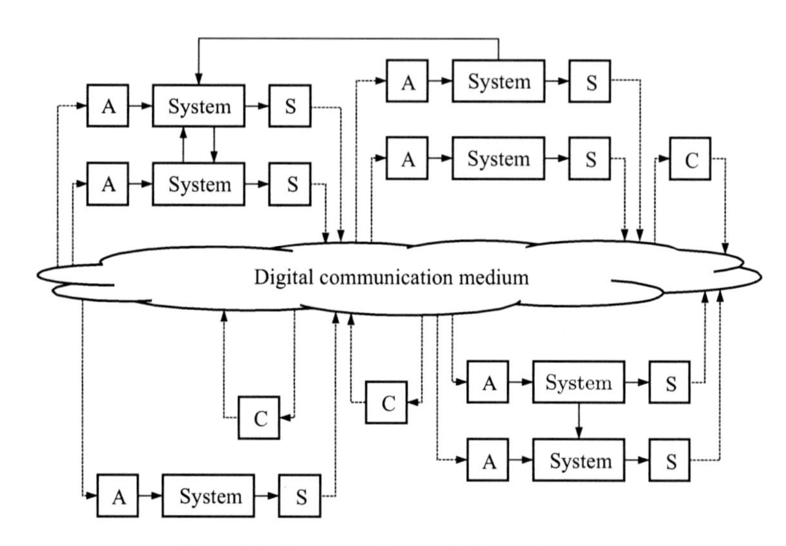

Figure 1.2.: Structure of networked control systems

However, the communication network has a considerable influence on the loop performance since its load affects the quality of service by inducing delays or packet losses which degrade the system performance or even cause the instability of the control loop. To avoid this situation, communication and control have to be investigated with respect to their interaction [93].

In this context, the analysis of event-based control as an alternative to discrete-time control has gained attention by considering event-based control as a suitable means to reduce the communication load of the network, see e.g. [8, 34, 40, 88, 117, 120]. The main aim to be reached by this feedback structure is the adaptation of the communication among the components of the feedback loop to the current system behaviour. In fact, by reducing the information exchange to the minimum communication that is necessary to ensure the required system performance, an overload of the digital communication network can be avoided.

Besides, networked control systems raise further questions with respect to event-based control, which concern e.g. the choice of suitable topologies, i.e. on which node should the components be located, and the sensitivity of event-based control towards delays or packet losses [59].

Low resolution sensors. A sensor-based event-driven control approach has been thoroughly studied in [105]. There, the starting point for the investigations has been the fact that in applications using low resolution sensors a periodic operation of the controller is not suitable because changes in the quantised measurements occur generally asynchronously in time. In this regard, the event-based nature of the sampling results from the measurement method.

Application examples are the encoder-based measuring of the angular position of a motor [55, 60], the use of capacitive level sensors for measuring the level of fluids in a tank (in e.g. [78]) and transportation systems where the longitudinal position of a vehicle is only known when certain markers are passed [43]. Evidently, the event-based nature of these sensors becomes more important when their resolutions are low.

Spaceflight. A further field, in which event-based control has been already applied, is the control of satellites, spacecrafts or space stations. A main concern is that the actuation elements, e.g. thrusters, operate in a fuel efficient way. For this purpose, instead of continuously adapting the controller signal, the thrusters impulsively unload their momentum only at certain time instants [98]. Thus, in comparison to the previous examples the event-based approach is based on a reasonable operation of the actuators and does not result from communication or sensor aspects.

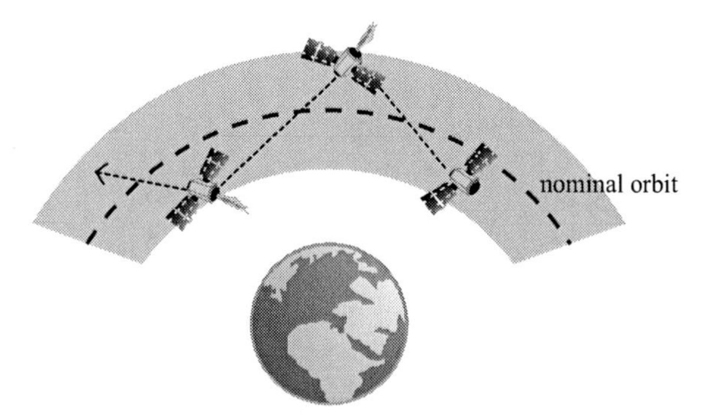

Figure 1.3.: Attitude control of satellites

A common task in astronautics concerns the stationkeeping problem or attitude control [64] (Fig. 1.3), which deals with the problem of keeping a satellite on a desired nominal orbit. Due to disturbances such as gravitation influences of the sun or the moon, which lead to a drift of the satellite, a precise path following is difficult to obtain. Instead, the satellite should be maintained within some torus centred about the nominal orbit. At certain times, the stationkeeping strategy has to compute the manoeuvres which drive the satellite back to an acceptable region within the torus. Generally, the problem consists of determining the required manoeuvres including the magnitude and the direction, and specifying the time instants at which the manoeuvres have to be executed.

In this context, reference [57] has proposed to use double gimballed solar arrays for attitude control in order to combine momentum control and optimal power generation. The paper showed that it is effective to control the solar arrays in an event-based way.

Further application fields. A particle accelerator, such as the Large Hadron Collider (LHC) [48], describes another suitable application example. Here, both the actuation (acceleration and path keeping) and the measuring do not occur continuously or after a fixed time interval but at certain discrete positions at the accelerator ring.

Furthermore, as stated in [27], event-based control compared to discrete-time control is closer in nature to the way a human behaves as a controller because a human usually performs manual control not until the measurement signal has deviated sufficiently enough from the desired setpoint [94].

Apart from these specific application examples, event-based control can be generally applied in all common control scenarios in which a bounded deviation from a setpoint or a reference trajectory can be tolerated.

1.1.3. Fundamental questions

The theoretical challenge of event-based control results from the following fact:

> As in the event-based control loop the sampling is generally non-periodic, the fundamental assumption of discrete-time control that the sampling times are equidistant, is violated.

Hence, for event-based control the well-known theory for discrete-time systems cannot be applied and, therefore, a new theory has to be developed. For this purpose, new models and methods for the description, the analysis and the design of the event-based control loop as well as new criterions for suitably evaluating its performance have to be found. Two of these aspects are emphasised next.

Analysis and design tasks. The main analysis and design tasks with respect to event-based control concern the choice of the event generator and the control input generator and can be summarised by three questions:

- At which time should the feedback link be closed?
- Which information should be sent at event times?
- What should the control input generator do between two consecutive event times with the information received at the event times?

Asymptotic stability vs. ultimate boundedness. In the event-based control loop an information exchange among the components is not triggered until the state $\boldsymbol{x}(t)$ of a system exceeds certain thresholds, e.g. a predefined deviation from the equilibrium point $\bar{\boldsymbol{x}}$. Hence, asymptotic stability cannot be ensured in general and a more appropriate stability definition for event-based control has to be considered which is given by *ultimate boundedness* (*practical stability*) [70]. The state $\boldsymbol{x}(t)$ of a system is called *ultimately bounded* if it can be driven into a predefined surrounding Ω_{t} of the equilibrium point $\bar{\boldsymbol{x}}$ and kept there for all future times $t > \bar{t}$ (see Section 2.1.2 for an illustration and a formal definition of ultimate boundedness).

1.2. Literature review

The crucial aspect for not applying event-based control ubiquitously is the absence of a fundamental and comprehensive theory. To fill this gap, a lot of experimental and analytical work has been done mainly in the recent years which is summarised in this section.

Early papers explicitly using the terminology *event-based control* (or *asynchronous control*) in their titles have been [27] and [55] which investigated event-based control by simulation and experiments. Reference [27] showed that an event-based implementation of a PID-controller allows a considerable reduction of the computational effort without appreciably degrading the system performance compared to a discrete-time implementation. The paper [55] experimentally showed the beneficial use of event-based control in the case of controlling motors with low resolution sensors.

In the recent years an increasing number of results on the analysis of event-based control of linear and nonlinear systems has been published. However, until now there is no uniform terminology describing this scheme [10]. It is called *event-based control* [28], *event-driven control* [56], *event-triggered control* [74], *Lebesgue control* [31], *deadband control* [97], *send-on-delta control* [121], *level-crossing control* [72], *asynchronous control* [55], *sporadic control* [61], *minimum attention control* [26], *interrupt-based control* [65], *need-based control* [41], *state-triggered control* [114] and *self-triggered control* [25].

Although the basic idea of all these approaches is the same, namely to sample a system only if certain event conditions are satisfied, their implementations may vary. The following paragraphs discuss the most important research directions of event-based control.

Deadband control. In [56, 105] event-based control has been investigated in the setting of linear systems of arbitrary dynamical order. The idea of the approach is to stick to a discrete-time state feedback with a fixed sampling period T_s only as long as the plant state $\boldsymbol{x}(t)$ is outside of a set (deadband) $\mathcal{B}$ of the state space around the origin and to ignore the sampling whenever the state is inside the set (left-hand side of Fig. 1.4). Based on the analysis of the corresponding discrete-time system and a corresponding piecewise linear representation, the state $\boldsymbol{x}(t)$ of the event-driven control loop has been proven to be ultimately bounded.

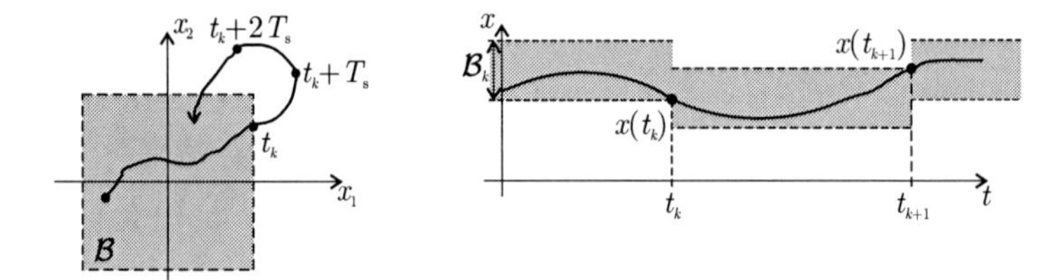

Figure 1.4.: Two deadband control schemes

Deadband control or send-on-delta control as proposed in [63, 97, 121] are similar schemes. Here, the deadband controller compares the state $x(t_k)$ of a first-order linear system previously sent over the network to the most recent value $x(t)$. If the absolute value of the difference is within the deadband $\mathcal{B}_k$ given by

$$\mathcal{B}_k = \{x \, : \, |x - x(t_k)| < \bar{e}\}$$

with the event threshold $\bar{e}$, no information is sent over the network (right-hand side of Fig. 1.4). If the difference reaches the boundary of this set, i.e. $|x(t) - x(t_k)| = \bar{e}$, the current measurement $x(t)$ is transmitted and a new deadband $\mathcal{B}_{k+1}$ is established around the value $x(t)$. In contrast to the previous approach, no periodic sampling is performed and the deadband varies depending on the last measurement. Although this scheme shows promising results in simulation, general analytical results are still missing.

A further approach has been studied in [46], where besides a deadband for the output $\boldsymbol{y}(t)$ an additional deadband for the input $\boldsymbol{u}(t)$ has been applied. The analysis uses an impulsive system description [52, 54] of the event-based control loop and linear matrix inequalities (LMIs)

[38] to show that the state of the event-based control loop is ultimately bounded and to derive a lower bound on the time interval between two consecutive event times.

Event-based control of linear stochastic systems. References [28, 30, 31, 58, 61] investigated the difference between a discrete-time control and an event-based control both of which use impulsive input signals. In the event-based control loop, a sampling is again triggered when the state of the system reaches a predefined threshold around the origin but, at event times, the state is now impulsively reset to the origin (Fig. 1.5). Using stochastic control theory [29], analytical results are obtained for first-order linear stochastic systems comparing the performance of the discrete-time and the event-based control loop. The results show that under certain circumstances the event-based control loop has even a better performance than the discrete-time loop because the communication is not invoked by an external clock but by the control error.

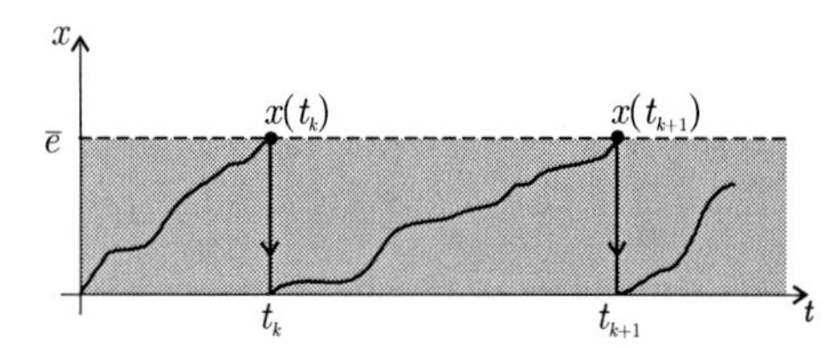

Figure 1.5.: Event-based control using impulsive inputs

In [42, 77, 90] the control and communication performance of event-based control of linear stochastic discrete-time systems of arbitrary dynamical order has been analysed in a similar way.

Lyapunov approaches to event-based control. There are some approaches in the literature which use Lyapunov-based methods [70] for determining the event times and analysing the event-based control loop. [113] showed for undisturbed nonlinear systems that by transmitting the state information $\boldsymbol{x}(t)$ to the controller only if the condition

$$\gamma(\|\boldsymbol{x}(t) - \boldsymbol{x}(t_k)\|) = \sigma\alpha(\|\boldsymbol{x}(t)\|)$$

is satisfied with $\sigma > 0$, asymptotic stability of the event-based controlled system can be guaranteed. Here, γ and α denote nondecreasing unbounded functions.

In a similar way, References [89, 124] proposed to generate events by explicitly evaluating Lyapunov functions. In these approaches, the control input $\boldsymbol{u}(t)$ is held constant between consecutive events and is recomputed if the Lyapunov function $V(\boldsymbol{x}(t_k), t)$ of the controlled system reaches the current value of a predefined performance function $S(\boldsymbol{x}(t_k), t)$, which describes the required behaviour of the control loop (Fig. 1.6).

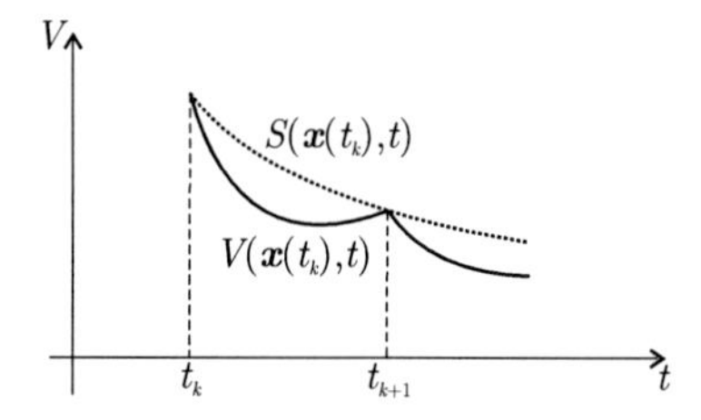

Figure 1.6.: Event triggering based on comparing Lyapunov functions

Level-triggered control. A different approach for limiting the information transmissions among the sensors, the controller and the actuators has been proposed e.g. in [39, 93, 99, 100, 130], where a quantisation of the signals is introduced. However, in this approach the reduction concerns the information contents rather than the communication as the information exchange is usually invoked in a discrete-time way.

References [4, 53, 60, 72] considered the combination of quantised control and event-based control by invoking a communication only if the state of the system or some other signal exceeds a certain quantisation interval (Fig. 1.7). Here, only quantised information is sent to the controller at event times. The quantisation itself can be fixed or time varying.

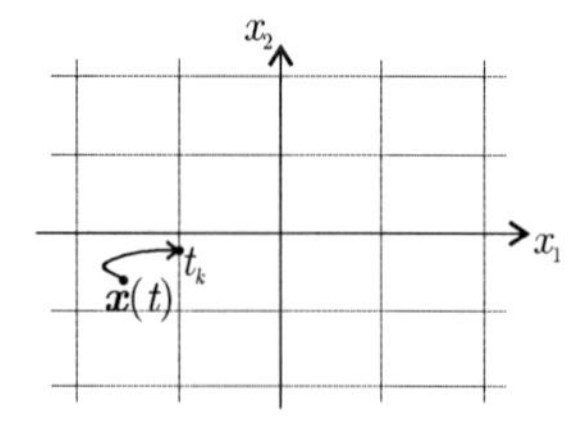

Figure 1.7.: Quantised event-based control

Self-triggered control. Self-triggered control [25, 86, 87, 122, 126] is closely related to event-based control but there exists a subtle difference. Instead of continuously monitoring an event condition, the idea of self-triggered control is to determine at event time t_k the next sampling time t_{k+1} in advance based on a function

$$t_{k+1} = h(\boldsymbol{x}(t_k), t_k).$$

This is illustrated in Fig. 1.8. To derive this relation, usually the evolution of certain Lyapunov and performance functions is investigated [89].

The main benefit of this approach lies in a reduced energy consumption of the sensor nodes because they can sleep until the next predicted sampling time. For this purpose, the sensor node has to include a scheduler which replaces the event generator (Fig. 1.1). A general problem of this scheme is the consideration of unknown effects, such as model uncertainties or unknown exogenous disturbances. To cope with all these effects conservative results have to be derived to guarantee the stability of the self-triggered control loop which may lead to relatively short sampling intervals in practice [128].

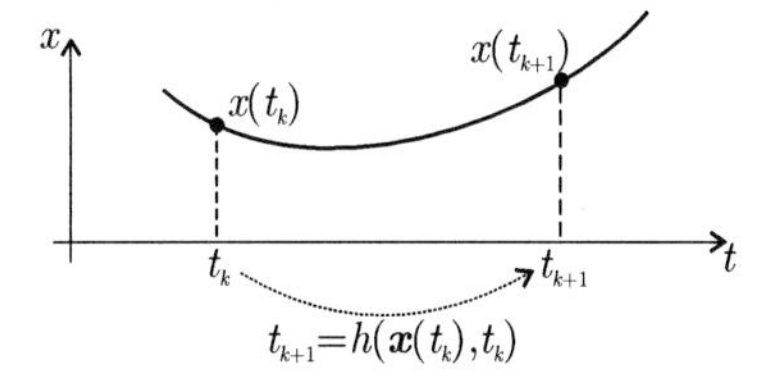

Figure 1.8.: Self-triggered control

Event-based control of multi-loop systems. All approaches mentioned have in common that they are concerned with single-loop systems, in which the sensor node and the actuator node are assumed to have access to all measurements and actuators of the plant. Especially in networked control systems, where both the sensors and the actuators may be widely distributed within the plant (Fig. 1.2), this situation is not always feasible.

Therefore, recent publications like [36, 40, 65, 88, 103, 125, 127, 129] investigated event-based control in the broader setting of multi-loop systems, where several loops are closed over the same communication network. For first-order linear stochastic systems, it has been shown in [40] that event-based control in combination with the communication protocol *Carrier Sense Multiple Access* (CSMA) guarantees a better performance than discrete-time control. However, other communication protocols may deteriorate the performance of the event-based control loop as shown in [36, 37].

In a similar setting references [44, 45, 47, 106] investigated the event-based control of multi-agent systems and event-based strategies for distributed model predictive control.

Critical remarks to the existing approaches. Surveying the existing approaches raises several critical comments:

- Some approaches like [56, 105] can be seen as extended discrete-time control schemes rather than event-based control since the event-based character of these schemes is marginal.
- A lot of approaches, e.g. [25, 72, 97, 125], do not consider any kind of uncertainties such as unknown exogenous disturbances which highly degrades their practical applicability.
- Almost all approaches use a zero-order hold for determining the continuous-time input $\boldsymbol{u}(t)$ between two consecutive events. Hence, doing nothing between two events is assumed to be the best means for obtaining an appropriate performance while reducing the communication. This consideration, however, is not convincing and serves rather as a means to simplify the analysis.

1.3. Contribution of this thesis

The approach discussed in this thesis and firstly proposed in the papers [3, 8] is called *event-based state feedback*. It has the following aim:

> The event-based state feedback should be capable to mimic a continuous-time state feedback with adjustable precision and adjustable communication.

As event-based control works in a closed-loop fashion only at event times t_k, when an information exchange is invoked, and in an open-loop structure between the event times, the underlying question how to determine the times t_k leads to the question under what condition is a feedback actually necessary. There are three general situations [79]:

- An unstable plant has to be stabilised.
- Feedback should allow the controller to deal with model uncertainties.
- Unknown disturbances have to be attenuated.

If none of these situations occur, the control input $\boldsymbol{u}(t)$ can be determined in a feedforward manner and no communication over the feedback link is necessary. In particular, if no disturbance occurs, a precise model is available and the output of a stable plant should be made to follow a command signal $\boldsymbol{w}(t)$, the control input $\boldsymbol{u}(t)$ can be fixed in a feedforward structure in terms of the command input $\boldsymbol{w}(t)$ [68].

Based on this consideration, the event-based state feedback has three novelties compared to the literature:

1. The control input generator is no longer a zero-order hold. Instead, it uses a feedforward control including a model of the continuous-time state-feedback loop to continuously adapt the input $\boldsymbol{u}(t)$ to a predicted state of the plant.

2. The event generator evaluates the current plant state in comparison with the state that the model of the continuous-time state-feedback loop has. Thus, an event does not indicate a large control error but a large deviation of the state of the event-based control loop from a desired reference state.

3. Both the event generator and the control input generator include a disturbance estimator, which, at event times, adapts the disturbance estimate used in both components to the current disturbance.

By designing the components of the event-based control loop according to these characteristics, the following properties hold:

- The approximation error between the state of the event-based state-feedback loop and the desirable state of the corresponding continuous-time state-feedback loop is bounded and can be arbitrarily adjusted by choosing the event threshold accordingly (Chapter 3, Theorem 3).

- The communication over the feedback link is bounded and depends explicitly on the disturbance magnitude. Likewise, it can be arbitrarily adjusted by the event threshold. However, by choosing the threshold, a tradeoff between a desired approximation accuracy and an acceptable communication has to be found (Chapter 3, Theorem 5).

The **contribution of this thesis** lies in the extension of the basic scheme. First, it complements the event-based state feedback by alternative methods for the analysis, new properties of the event-based control loop and improvements of the existing components, i.e.

- a **Lyapunov-based stability analysis** (Section 3.4.2, Theorem 4),

- an investigation of the **setpoint tracking properties** of the event-based state-feedback loop (Section 3.4.5, Theorem 6) and an improvement of these properties by incorporating a **dynamical controller** in the control input generator (Section 4.3, Theorem 12),
- an **improved disturbance estimation** (Section 4.4),
- a **common basis for comparing** event-based state feedback and discrete-time state feedback (Section 3.5, Theorem 8), and
- a **discrete-time realisation** of the control input generator and the event generator which is motivated by the practical implementation of these components (Section 4.5, Theorems 13, 14).

Second, it investigates the consequences of non-ideal plant and communication properties on the event-based control, and their compensation. This includes the elaboration of

- suitable strategies for compensating **imperfect plant information** such as model uncertainties and non-measurable state variables (Sections 3.4.6, 4.2, 4.4, Theorems 7, 9), and
- appropriate modifications of the event-based control loop to overcome **imperfect communication properties** such as delays, packet losses and data-rate constraints (Chapter 5, Theorems 15, 16).

In all these cases, it is shown that the event-based state-feedback loop remains stable and an upper bound on the approximation error with respect to the behaviour of the continuous-time state-feedback loop can be determined. However, when considering plant uncertainties or communication imperfections, the performance of the event-based control loop may deteriorate in terms of the guaranteed approximation accuracy. As expected, these non-ideal properties generally lead to an increased information exchange over the feedback link in the control loop but the communication remains bounded as well (Theorems 10, 17, 18). Note that the analysis in this thesis solely concentrates on systems in which the sensor node and the actuator node are assumed to have access to all measurements and actuators of the plant (Fig. 1.1).

Third, a thermofluid process is defined which is used throughout the thesis to evaluate the elaborated concepts in **simulations and experiments**. The experimental evaluation (Chapter 6) validates that

- the proposed schemes are **robust** against severe model uncertainties, and
- plant uncertainties and communication imperfections rather increase the communication than deteriorate the performance of the event-based control loop. However, the communication is generally significantly lower than in a discrete-time state feedback.

1.4. Structure of this thesis

Chapter 2 introduces the mathematical notations and definitions which are used throughout this thesis. Event-based sampling and discrete-time sampling are compared to allow a better distinction between their respective notations. A thermofluid process is described which is used as a running example in order to illustrate the elaborated concepts and main results.

Chapter 3 presents the basic concept of the event-based state feedback and general methods for the analysis of the behaviour of the event-based control loop with respect to its stability and communication properties. These methods provide the fundament for the analytical investigations in the subsequent chapters.

Chapter 4 extends the event-based state feedback with the aim to improve its practical applicability. After presenting an output-feedback scheme including a state observer, the effects of using a dynamical controller and alternative disturbance estimators on the event-based control loop are investigated. Finally, a discrete-time realisation of the event generator and the control input generator is considered.

Chapter 5 investigates the influence of a non-ideal communication channel on the event-based control scheme and proposes strategies to suitably deal with communication imperfections. The analysis concentrates on three effects: communication delays, packet losses, and data-rate constraints.

Chapter 6 presents the experimental evaluation of the event-based level and temperature control of the thermofluid process introduced in Chapter 2. It additionally includes comments to the practical implementation of the event-based control loop.

Chapter 7 summarises and concludes the thesis and presents possible directions for future research.

2. Preliminaries

This chapter summarises the general notations used in this thesis. A detailed list of symbols can be found in Appendix C. Moreover, this chapter introduces a thermofluid process which serves as a running example in the following chapters.

2.1. Notations

2.1.1. General definitions

Throughout this thesis a scalar is denoted by italic letters ($x \in \mathbb{R}$), a vector by bold italic letters ($\boldsymbol{x} \in \mathbb{R}^n$), a matrix by upper-case bold italic letters ($\boldsymbol{A} \in \mathbb{R}^{n \times n}$) and a signal at time $t \in \mathbb{R}_+$ by $\boldsymbol{x}(t)$, where $\boldsymbol{x}_0$ is defined as the initial signal value at time $t = 0$. $\boldsymbol{x}'$ and $\boldsymbol{A}'$ denote the transpose of a vector or matrix, respectively. $\boldsymbol{H}^{-1}$ denotes the inverse of a square matrix $\boldsymbol{H} \in \mathbb{R}^{n \times n}$ and $\boldsymbol{H}^+$ is the pseudoinverse of a non-square matrix $\boldsymbol{H} \in \mathbb{R}^{m \times l}$ which can be determined according to

$$\boldsymbol{H}^+ = \begin{cases} (\boldsymbol{H}'\boldsymbol{H})^{-1}\boldsymbol{H}', & \text{for } m \geq l \\ \boldsymbol{H}'(\boldsymbol{H}\boldsymbol{H}')^{-1}, & \text{for } l \geq m, \end{cases} \tag{2.1}$$

if the matrix $\boldsymbol{H}$ has full rank [35], i.e.

$$\text{Rank}(\boldsymbol{H}) = \begin{cases} l, & \text{for } m \geq l \\ m, & \text{for } l \geq m. \end{cases}$$

For $m \geq l$, $\boldsymbol{H}^+$ is the left inverse of $\boldsymbol{H}$ ($\boldsymbol{H}^+\boldsymbol{H} = \boldsymbol{I}_l$), whereas, for $l \geq m$, $\boldsymbol{H}^+$ is the right inverse of $\boldsymbol{H}$ ($\boldsymbol{H}\boldsymbol{H}^+ = \boldsymbol{I}_m$). $\boldsymbol{I}_n \in \mathbb{R}^{n \times n}$ denotes the identity matrix of size n. $\boldsymbol{O}$ and 0 denote a zero matrix or a zero vector of appropriate dimension, respectively.

The absolute value is denoted by $|x|$. The notations $\|\boldsymbol{x}\|$ and $\|\boldsymbol{A}\|$ are used to denote an arbitrary vector or matrix norm according to

$$\|\boldsymbol{x}\|_p := \left(\sum_{i=1}^{n} |x_i|^p\right)^{\frac{1}{p}}$$
$$\|\boldsymbol{A}\|_p := \sup_{\boldsymbol{x}\neq\boldsymbol{0}} \frac{\|\boldsymbol{A}\boldsymbol{x}\|_p}{\|\boldsymbol{x}\|_p}$$

and $\|\boldsymbol{x}\|_\infty$ is the supremum norm of a vector according to

$$\|\boldsymbol{x}\|_\infty := \max_{i\in\{1,\dots,n\}} |x_i|$$

[50]. Moreover, $\|\boldsymbol{x}(t)\|$ denotes an arbitrary vector norm at time t. Eigenvalues of a square matrix are denoted by λ, where $\lambda_{\min}(\boldsymbol{A})$ and $\lambda_{\max}(\boldsymbol{A})$ are the minimum or maximum eigenvalue of the matrix $\boldsymbol{A}$, respectively. A set is denoted by $\Omega \subseteq \mathbb{R}^n$ and its boundary is given by $\partial\Omega$.

A continuous-time linear time-invariant system is given by

$$\dot{\boldsymbol{x}}(t) = \boldsymbol{A}\boldsymbol{x}(t) + \boldsymbol{B}\boldsymbol{u}(t) + \boldsymbol{E}\boldsymbol{d}(t), \qquad \boldsymbol{x}(0) = \boldsymbol{x}_0 \tag{2.2}$$
$$\boldsymbol{y}(t) = \boldsymbol{C}\boldsymbol{x}(t), \tag{2.3}$$

where $\boldsymbol{x} \in \mathbb{R}^n$ denotes the state of the system with the initial value $\boldsymbol{x}_0$, $\boldsymbol{u} \in \mathbb{R}^m$ and $\boldsymbol{y} \in \mathbb{R}^r$ are the inputs or measured outputs, respectively, and $\boldsymbol{d} \in \mathbb{R}^l$ represents exogenous disturbances. $\boldsymbol{A} \in \mathbb{R}^{n\times n}$, $\boldsymbol{B} \in \mathbb{R}^{n\times m}$, $\boldsymbol{E} \in \mathbb{R}^{n\times l}$ and $\boldsymbol{C} \in \mathbb{R}^{r\times n}$ are real matrices.

Definition 1. [70] *A square matrix $\boldsymbol{A}$ is called Hurwitz if each eigenvalue*

$$\lambda_i\{\boldsymbol{A}\}, \quad i \in 1, 2, \dots, n$$

of $\boldsymbol{A}$ has strictly negative real part.

Theorem 1. [70] *The system* (2.2), (2.3) *with zero inputs ($\boldsymbol{u}(t) = \boldsymbol{0}$ and $\boldsymbol{d}(t) = \boldsymbol{0}$ for all times $t \geq 0$) is asymptotically stable, i.e.*

$$\lim_{t\to\infty} \|\boldsymbol{x}(t)\| = 0,$$

if the matrix $\boldsymbol{A}$ is Hurwitz.

2.1.2. Ultimate boundedness

Section 1.1.3 raised that claiming asymptotic stability in the context of event-based control is inadequate. This likewise holds for the continuous-time system (2.2), (2.3) affected by some nonzero exogenous signals $\boldsymbol{u}(t)$ and $\boldsymbol{d}(t)$ which are bounded according to

$$\begin{aligned} \|\boldsymbol{u}(t)\| &\leq u_{\max} \\ \|\boldsymbol{d}(t)\| &\leq d_{\max}. \end{aligned}$$

A more appropriate stability definition is given by ultimate boundedness which is illustrated in Fig. 2.1 and defined next.

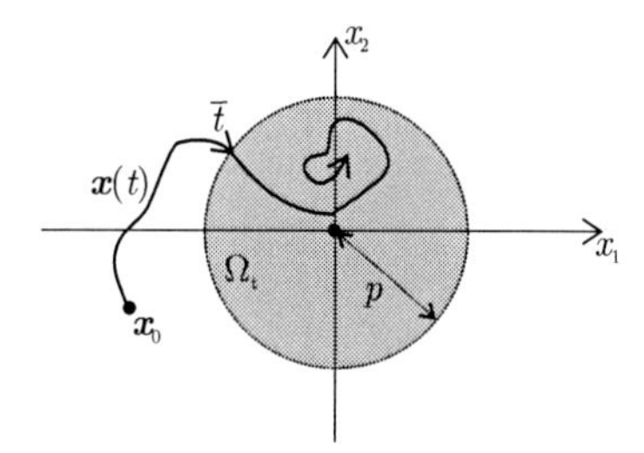

Figure 2.1.: Ultimate boundedness

Definition 2. **[70]** *The solution $\boldsymbol{x}(t)$ of the continuous-time system* (2.2), (2.3) *is globally uniformly ultimately bounded (GUUB) if for every $\boldsymbol{x}(0) \in \mathbb{R}^n$ there exists a positive constant p and a time $\bar{t}$ such that holds:*

$$\boldsymbol{x}(t) \in \Omega_{\mathrm{t}} = \{\boldsymbol{x} : \|\boldsymbol{x}\| \leq p\}, \quad \forall t \geq \bar{t}. \tag{2.4}$$

Definition 3. *The continuous-time system* (2.2), (2.3) *is called ultimately bounded or stable if its state is GUUB.*

Theorem 2. **[70]** *The state $\boldsymbol{x}(t)$ of the continuous-time system* (2.2), (2.3) *is GUUB if the matrix $\boldsymbol{A}$ is Hurwitz and the input $\mathbf{u}(t)$ and the disturbance $\mathbf{d}(t)$ are bounded.*

In literature there are further equivalent stability definitions such as input-to-state stability [70, 86, 109] or practical stability [49, 73] which all refer to the same situation, namely that the state should be brought into a desired region Ω_{t} and kept there for all future times despite exogenous disturbances.

2.1.3. Event-based sampling and discrete-time sampling

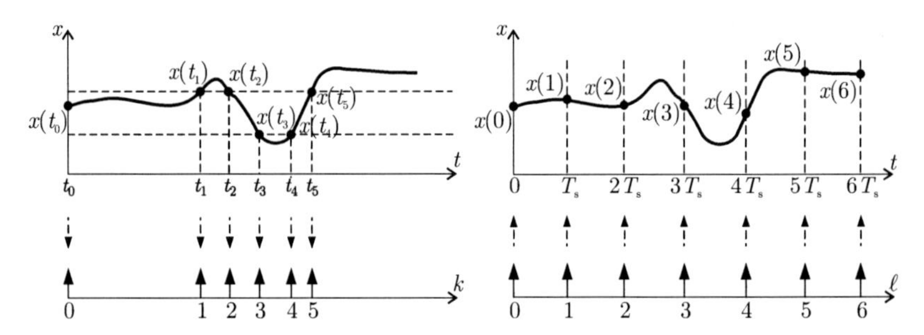

Figure 2.2.: Event-based sampling and discrete-time sampling

As mentioned in the introduction, the sampling in the event-based control loop differs fundamentally from the sampling in the discrete-time control loop. Figure 2.2 illustrates the difference between the event times t_k with the event counter $k \in \{0, 1, ...\}$ (left-hand side of the figure) and the discrete-time instants $\ell \in \{0, 1, ...\}$ which specify the corresponding sampling times according to ℓT_s with the fixed sampling period T_s (right-hand side of the figure).

In contrast to the discrete-time control, in the event-based control loop the sampling is not invoked by an external clock but by the system behaviour exceeding certain bounds. Hence, in time intervals in which the system behaviour does not vary significantly a sampling occurs rarely. However, in time intervals, in which disturbances force strong changes in the system behaviour, an increased number of samplings is invoked by the event generator.

Definition 4. **[46]** *The time interval T_k between two consecutive event times*

$$T_k = t_{k+1} - t_k, \quad k = 0, 1, 2, ... \tag{2.5}$$

is called the inter-event time.

2.2. Running example: A thermofluid process

Figure 2.3 shows the experimental set-up of a thermofluid process which is used as a running example throughout this thesis in order to illustrate the concepts and to quantitatively evaluate the theoretical results. Experimental results are collectively presented in Chapter 6. The thermofluid process considered has been introduced in [14] and used e.g. in [1, 3, 4, 5, 15, 53] as an illustrative example.

The central component of the process is the cylindrical batch reactor TB in which the level $l_{\mathrm{TB}}(t)$ and the temperature $\vartheta_{\mathrm{TB}}(t)$ of a fluid should be held on prescribed operating points $\bar{l}_{\mathrm{TB}}$ and $\bar{\vartheta}_{\mathrm{TB}}$.

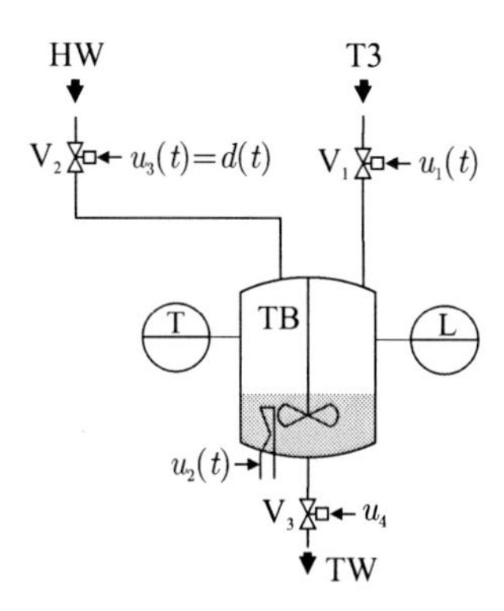

Figure 2.3.: Experimental set-up of the thermofluid process

In the reactor, both the level $l_{\mathrm{TB}}(t)$ and the temperature $\vartheta_{\mathrm{TB}}(t)$ can be continuously measured. The valve angle $u_1(t) \in [0,1]$ which controls the inflow into tank TB and the signal $u_2(t) \in [0,6]$ which corresponds to the heater included in the tank can be continuously adjusted and are used as inputs: $\boldsymbol{u}(t) = (u_1(t) \;\; u_2(t))'$. In contrast, the valve angle $u_3(t) \in [0,1]$ providing an additional inflow into tank TB is used as an unknown disturbance: $d(t) = u_3(t)$. A continuous outflow of the reactor TB is realised over the valve angle u_4 which is held in a fixed position $(u_4 = 0.2)$ during the process.

Defining the state as

$$\boldsymbol{x}(t) = \begin{pmatrix} l_{\mathrm{TB}}(t) \\ \vartheta_{\mathrm{TB}}(t) \end{pmatrix} = \begin{pmatrix} x_1(t) \\ x_2(t) \end{pmatrix},$$

the process is described by a nonlinear model whose state-space representation can be found in Appendix A, page 169. As the following investigations are concerned with continuous-time linear time-invariant systems of the form (2.2), (2.3), the nonlinear model has been linearised around the operating point

$$\bar{\boldsymbol{x}} = \begin{pmatrix} \bar{x}_1 \\ \bar{x}_2 \end{pmatrix} = \begin{pmatrix} 40\,\mathrm{cm} \\ 313\,\mathrm{K} \end{pmatrix} \tag{2.6}$$

$$\bar{\boldsymbol{u}} = \begin{pmatrix} \bar{u}_1 \\ \bar{u}_2 \end{pmatrix} = \begin{pmatrix} 0.36 \\ 1.65 \end{pmatrix},$$

which yields the linearised model

$$\delta\dot{\boldsymbol{x}}(t) = 10^{-3}\begin{pmatrix} -0.8 & 0 \\ -1\cdot 10^{-7} & -1.7 \end{pmatrix}\delta\boldsymbol{x}(t) + 10^{-3}\begin{pmatrix} 211 & 0 \\ -108 & 20 \end{pmatrix}\delta\boldsymbol{u}(t) \tag{2.7}$$

$$+\ 10^{-3}\begin{pmatrix} 148 \\ -80 \end{pmatrix}\delta d(t), \quad \delta\boldsymbol{x}(0) = \delta\boldsymbol{x}_0$$

$$\delta\boldsymbol{y}(t) = \begin{pmatrix} 1 & 0 \\ 0 & 1 \end{pmatrix}\delta\boldsymbol{x}(t) \tag{2.8}$$

with $\delta\boldsymbol{x}(t) \triangleq \boldsymbol{x}(t) - \bar{\boldsymbol{x}}$, $\delta\boldsymbol{u}(t) \triangleq \boldsymbol{u}(t) - \bar{\boldsymbol{u}}$ and $\delta d(t) \triangleq d(t) - \bar{d}\,(\bar{d} = 0)$. The unit of the level x_1 is cm and the unit of the temperature x_2 is K. Henceforth, the prefix δ is neglected as the linearised model (2.7), (2.8) is always considered.

3. A state-feedback approach to event-based control

This chapter presents a state-feedback approach to event-based control. Stability as well as communication properties of the event-based control loop are derived and compared to respective properties of the discrete-time state-feedback loop.

3.1. Assumptions

This chapter deals with the basic concept of the event-based state feedback proposed in [8, 81] and complements this scheme by investigating

- an alternative Lyapunov-based stability analysis (Section 3.4.2),
- the setpoint tracking properties of the event-based state feedback (Section 3.4.5),
- the robustness of the event-based control loop (Section 3.4.6), and
- the relation between event-based state feedback and discrete-time state feedback (Section 3.5).

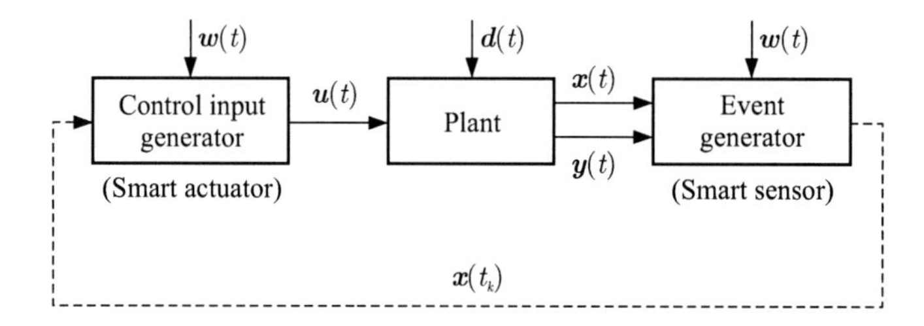

Figure 3.1.: Simplified event-based control loop

In the sequel of this thesis the event-based control loop shown in Fig. 1.1 is simplified as depicted in Fig. 3.1. Here, the controller is incorporated in the control input generator, because in the following analysis it is not important how the task to determine the continuous-time input signal $\boldsymbol{u}(t)$ is distributed among these components. Again, the dashed line in the figure indicates that the link is only used at event times t_k ($k = 0, 1, 2, ...$), whereas the links drawn by solid lines are used continuously.

The plant is given by the linear state-space model

$$\dot{\boldsymbol{x}}(t) = \boldsymbol{A}\boldsymbol{x}(t) + \boldsymbol{B}\boldsymbol{u}(t) + \boldsymbol{E}\boldsymbol{d}(t), \qquad \boldsymbol{x}(0) = \boldsymbol{x}_0 \tag{3.1}$$

$$\boldsymbol{y}(t) = \boldsymbol{C}\boldsymbol{x}(t), \tag{3.2}$$

where the pair $(\boldsymbol{A}, \ \boldsymbol{B})$ is assumed to be controllable and the disturbance $\boldsymbol{d}(t)$ is bounded according to

$$\|\boldsymbol{d}(t)\| \leq d_{\text{max}}. \tag{3.3}$$

Note that these assumptions hold throughout the overall thesis. It is further assumed that

- the plant dynamics are accurately known,
- the state $\boldsymbol{x}(t)$ is measurable, and
- the information exchange between the event generator and the control input generator is instantaneous and imposes no restrictions on the information to be sent at event times.

Hence, the reason to communicate information via the dashed arrows in Fig. 3.1 is primarily given by the situation that the disturbance $\boldsymbol{d}(t)$ causes an intolerable behaviour of the control output $\boldsymbol{y}(t)$ or the plant state $\boldsymbol{x}(t)$.

Besides, the latter three assumptions will be relaxed later in this thesis by investigating the consequences of model uncertainties (Section 3.4.6), non-measurable state variables (Section 4.2) and imperfect communication links (Chapter 5).

The analysis in this chapter will show that by using exponential input signals $\boldsymbol{u}(t)$ provided by the control input generator (Section 3.3.1), by sending the state information $\boldsymbol{x}(t_k)$ at event times ($\boldsymbol{s}(t_k) = \boldsymbol{x}(t_k)$) and by determining the event times based on a comparison of the actual plant behaviour and the desired behaviour of a reference system (Section 3.3.3), the event-based state feedback has the following properties:

- The state $\boldsymbol{x}(t)$ of the event-based state-feedback loop is ultimately bounded in the sense that it remains, for all times t, in a bounded surrounding Ω_e of the aspired state $\boldsymbol{x}_\text{CT}(t)$ of the continuous-time state-feedback loop (Theorem 3).

- The communication over the feedback link in the event-based control loop is bounded and depends explicitly on the disturbance $\boldsymbol{d}(t)$ (Theorem 5).
- Both the approximation accuracy and the minimum inter-event time of the event-based control loop can be arbitrarily adjusted by changing the threshold parameter $\bar{e}$ in order to adapt the event-based state-feedback loop to the requested needs.

3.2. Continuous-time state feedback

3.2.1. Model

This section describes a continuous-time state-feedback loop, which will be used later as the reference system whose behaviour should be approximated by the event-based state-feedback loop.

The plant (3.1), (3.2) together with the state feedback

$$\boldsymbol{u}(t) = -\boldsymbol{K}\boldsymbol{x}(t) + \boldsymbol{V}\boldsymbol{w}(t), \tag{3.4}$$

where $\boldsymbol{w}(t) \in \mathbb{R}^r$ denotes the reference signal to be followed by the plant output $\boldsymbol{y}(t)$, yields the continuous-time closed-loop system

$$\dot{\boldsymbol{x}}_{\mathrm{CT}}(t) = \underbrace{(\boldsymbol{A} - \boldsymbol{B}\boldsymbol{K})}_{\bar{\boldsymbol{A}}}\boldsymbol{x}_{\mathrm{CT}}(t) + \boldsymbol{E}\boldsymbol{d}(t) + \boldsymbol{B}\boldsymbol{V}\boldsymbol{w}(t), \quad \boldsymbol{x}_{\mathrm{CT}}(0) = \boldsymbol{x}_0 \tag{3.5}$$

$$\boldsymbol{y}_{\mathrm{CT}}(t) = \boldsymbol{C}\boldsymbol{x}_{\mathrm{CT}}(t). \tag{3.6}$$

Here, the state and the output are associated with the index "CT" in order to distinguish them from the state and the output of the event-based control loop considered later. The state-feedback matrix $\boldsymbol{K}$ is assumed to be chosen so that the matrix $\bar{\boldsymbol{A}}$ is Hurwitz and the closed-loop system has a desired disturbance attenuation property.

> As $\bar{\boldsymbol{A}}$ is Hurwitz and the disturbance $\boldsymbol{d}(t)$ is assumed to be bounded according to Eq. (3.3), the state $\boldsymbol{x}_{\mathrm{CT}}(t)$ of the continuous-time state-feedback loop is *GUUB* for a bounded command input $\boldsymbol{w}(t)$ according to Theorem 2.

The matrix $\boldsymbol{V}$ is given by

$$\boldsymbol{V} = -\left(\boldsymbol{C}\bar{\boldsymbol{A}}^{-1}\boldsymbol{B}\right)^{+}, \tag{3.7}$$

where $(.)^+$ denotes the pseudoinverse which can be determined according to the formula (2.1). It exists if the matrix $\boldsymbol{C}\bar{\boldsymbol{A}}^{-1}\boldsymbol{B}$ has full rank.

For $m \geq r$, Eq. (3.7) ensures setpoint tracking of the undisturbed closed-loop system ($\boldsymbol{d}(t) = \boldsymbol{0}$) for a constant command signal $\boldsymbol{w}(t) = \bar{\boldsymbol{w}}$ [83]:

$$\lim_{t \to \infty} \|\boldsymbol{y}_{\mathrm{CT}}(t) - \bar{\boldsymbol{w}}\| = 0.$$

Henceforth, either the controller (3.4) or the simplified controller

$$\boldsymbol{u}(t) = -\boldsymbol{K}\boldsymbol{x}(t) \tag{3.8}$$

($\boldsymbol{w}(t) = \boldsymbol{0}$) is considered. The simplified controller is used in situations in which setpoint tracking is not of primary interest or makes the analysis unnecessarily complex.

Note that any state $\boldsymbol{x} \neq \boldsymbol{0}$ of the undisturbed plant (3.1), (3.2) can be moved into the origin by transforming the plant model accordingly [70]. Therefore, any setpoint $\boldsymbol{w}(t) \neq \boldsymbol{0}$ can also be reached by the undisturbed control loop with the transformed plant model and controller (3.8) if the resulting loop is asymptotically stable.

3.2.2. Communication structure

To implement the controller (3.4), the state $\boldsymbol{x}(t)$ has to be communicated continuously from the sensors towards the controller, which immediately sends $\boldsymbol{u}(t) = -\boldsymbol{K}\boldsymbol{x}(t) + \boldsymbol{V}\boldsymbol{w}(t)$ towards the actuators. This continuous information transfer is possible only if the control loop is implemented by analogue hardware.

In a discrete-time version, the state is measured at the time instants ℓT_{s} (Section 2.1.3). The feedback loop is closed at these times and the actuators receive the information

$$\boldsymbol{u}(\ell T_{\mathrm{s}}) = -\boldsymbol{K}\boldsymbol{x}(\ell T_{\mathrm{s}}) + \boldsymbol{V}\boldsymbol{w}(\ell T_{\mathrm{s}})$$

that is used for the whole sampling period:

$$\boldsymbol{u}(t) = \boldsymbol{u}(\ell T_{\mathrm{s}}) \text{ for } \ell T_{\mathrm{s}} \leq t < (\ell + 1)T_{\mathrm{s}}. \tag{3.9}$$

3.2.3. Behaviour of the continuous-time state-feedback loop

For the event-based controller, the continuous-time state-feedback loop (3.5), (3.6) rather than the discrete-time version should be used for comparison because sampling brings about a deterioration of the loop performance. The behaviour of the continuous-time state-feedback loop (3.5), (3.6) is given by the state trajectory

$$\boldsymbol{x}_{\mathrm{CT}}(t) = \mathrm{e}^{\bar{\boldsymbol{A}}t}\boldsymbol{x}_0 + \int_0^t \mathrm{e}^{\bar{\boldsymbol{A}}(t-\alpha)}\boldsymbol{E}\boldsymbol{d}(\alpha)\,\mathrm{d}\alpha + \int_0^t \mathrm{e}^{\bar{\boldsymbol{A}}(t-\alpha)}\boldsymbol{B}\boldsymbol{V}\boldsymbol{w}(\alpha)\,\mathrm{d}\alpha$$

and the output trajectory

$$\boldsymbol{y}_{\mathrm{CT}}(t) = \boldsymbol{C}\mathrm{e}^{\bar{\boldsymbol{A}}t}\boldsymbol{x}_0 + \int_0^t \boldsymbol{C}\mathrm{e}^{\bar{\boldsymbol{A}}(t-\alpha)}\boldsymbol{E}\boldsymbol{d}(\alpha)\,\mathrm{d}\alpha + \int_0^t \boldsymbol{C}\mathrm{e}^{\bar{\boldsymbol{A}}(t-\alpha)}\boldsymbol{B}\boldsymbol{V}\boldsymbol{w}(\alpha)\,\mathrm{d}\alpha.$$

The control input generated by the continuous-time state-feedback controller is given by

$$\begin{aligned}\boldsymbol{u}(t) &= -\boldsymbol{K}\mathrm{e}^{\bar{\boldsymbol{A}}t}\boldsymbol{x}_0 - \int_0^t \boldsymbol{K}\mathrm{e}^{\bar{\boldsymbol{A}}(t-\alpha)}\boldsymbol{E}\boldsymbol{d}(\alpha)\,\mathrm{d}\alpha \\ &\quad - \int_0^t \boldsymbol{K}\mathrm{e}^{\bar{\boldsymbol{A}}(t-\alpha)}\boldsymbol{B}\boldsymbol{V}\boldsymbol{w}(\alpha)\,\mathrm{d}\alpha + \boldsymbol{V}\boldsymbol{w}(t).\end{aligned}$$

This equation shows that the input $\boldsymbol{u}(t)$ does not only depend upon the initial state $\boldsymbol{x}_0$ and the command signal $\boldsymbol{w}(t)$, but also on the disturbance input $\boldsymbol{d}(t)$.

In the more general setting of event-based control, this fact is important. If at time t_k the state $\boldsymbol{x}(t_k)$ is communicated to the control input generator, the control input generator is able to determine the same control input $\boldsymbol{u}(t_k) = -\boldsymbol{K}\boldsymbol{x}(t_k) + \boldsymbol{V}\boldsymbol{w}(t_k)$ as a continuous-time state-feedback controller. However, in order to be able to determine the control input for some future times $t \geq t_k$, the control input generator has to know the disturbance $\boldsymbol{d}(t)$ for $t > t_k$:

$$\begin{aligned}\boldsymbol{u}(t) &= -\boldsymbol{K}\mathrm{e}^{\bar{\boldsymbol{A}}(t-t_k)}\boldsymbol{x}(t_k) - \int_{t_k}^t \boldsymbol{K}\mathrm{e}^{\bar{\boldsymbol{A}}(t-\alpha)}\boldsymbol{E}\boldsymbol{d}(\alpha)\,\mathrm{d}\alpha \\ &\quad - \int_{t_k}^t \boldsymbol{K}\mathrm{e}^{\bar{\boldsymbol{A}}(t-\alpha)}\boldsymbol{B}\boldsymbol{V}\boldsymbol{w}(\alpha)\,\mathrm{d}\alpha + \boldsymbol{V}\boldsymbol{w}(t), \quad t \geq t_k. \end{aligned} \tag{3.10}$$

Note that the command signal $\boldsymbol{w}(t)$ is assumed to be continuously available at the control input generator (Fig. 3.1).

This analysis shows two important facts:

- Continuous-time state-feedback control (3.4) gets the information about the current disturbance implicitly due to the continuous communication of the current state $\boldsymbol{x}(t)$.

- Any feedback without continuous communication has to make assumptions about the disturbance to be attenuated. Unless the disturbance is measurable, any discontinuous feedback cannot have the same performance as the feedback loop with continuous communication.

The second fact is also relevant for discrete-time control, where it is usually assumed that the control input at time ℓT_s is the appropriate input for the whole time period until the next sampling time (Eq. (3.9)). Only if this assumption is satisfied it is reasonable to determine the control input for the whole sampling period by a zero-order hold.

The main idea of the event-based state-feedback approach discussed in the following is to replace the continuous-time state feedback (3.4) by an event-based controller so that the state $\boldsymbol{x}(t)$ of the event-based state-feedback loop remains, for all times t, in a surrounding $\Omega_e(\boldsymbol{x}_{CT}(t))$ of the state $\boldsymbol{x}_{CT}(t)$ of the continuous-time state-feedback loop (3.5), (3.6).

3.3. Event-based state feedback

3.3.1. Model

In the event-based state-feedback loop the control input generator determines the input $\boldsymbol{u}(t)$ for the time $t \geq t_k$ in terms of the state information $\boldsymbol{x}(t_k)$ received at the event time t_k.

A direct consequence of the analysis in the preceding section is the fact that for the time $t \geq t_k$ the plant

$$\begin{aligned} \dot{\boldsymbol{x}}(t) &= \boldsymbol{A}\boldsymbol{x}(t) + \boldsymbol{B}\boldsymbol{u}(t) + \boldsymbol{E}\boldsymbol{d}(t), \qquad \boldsymbol{x}(t_k) = \boldsymbol{x}_k \\ \boldsymbol{y}(t) &= \boldsymbol{C}\boldsymbol{x}(t) \end{aligned}$$

with the control input (3.10) behaves exactly like the continuous-time control loop (3.5), (3.6). That is, if the control input generator uses Eq. (3.10) to determine the control input for the time $t \geq t_k$, then the best possible performance is obtained. The equation shows that the state $\boldsymbol{x}(t_k)$ has to be measured and communicated to the control input generator, which likewise holds in the continuous-time state-feedback loop. However, the equation also shows that this control law necessitates the knowledge about the disturbance $\boldsymbol{d}(t)$ for the time interval $[t_k, t_{k+1})$. As the disturbance is generally unknown, an assumption concerning its behaviour has to be made to get an implementable control law.

In the following, it is assumed that the disturbance is constant

$$\boldsymbol{d}(t) = \hat{\boldsymbol{d}}_k \quad \text{for} \quad t \geq t_k$$

and the magnitude $\hat{\boldsymbol{d}}_k$ is known. Hence, the control input generator uses the equation

$$\begin{aligned}\boldsymbol{u}(t) &= -\boldsymbol{K}\mathrm{e}^{\bar{\boldsymbol{A}}(t-t_k)}\boldsymbol{x}(t_k) - \boldsymbol{K}\bar{\boldsymbol{A}}^{-1}\left(\mathrm{e}^{\bar{\boldsymbol{A}}(t-t_k)} - \boldsymbol{I}_n\right)\boldsymbol{E}\hat{\boldsymbol{d}}_k \\ &\quad - \int_{t_k}^{t} \boldsymbol{K}\mathrm{e}^{\bar{\boldsymbol{A}}(t-\alpha)}\boldsymbol{B}\boldsymbol{V}\boldsymbol{w}(\alpha)\,\mathrm{d}\alpha + \boldsymbol{V}\boldsymbol{w}(t), \quad t \geq t_k \end{aligned} \tag{3.11}$$

which directly follows from Eq. (3.10) for constant disturbances, until it gets the next information $\boldsymbol{x}(t_{k+1})$. For $t = t_k$, the control input $\boldsymbol{u}(t_k) = -\boldsymbol{K}\boldsymbol{x}(t_k) + \boldsymbol{V}\boldsymbol{w}(t_k)$ is the same as in the continuous-time state-feedback loop. For $t > t_k$, the input signal $\boldsymbol{u}(t)$ depends upon the disturbance estimate $\hat{\boldsymbol{d}}_k$. Note that discrete-time control simply holds the control input constant at the last value and does not use any further information about the disturbance to be attenuated.

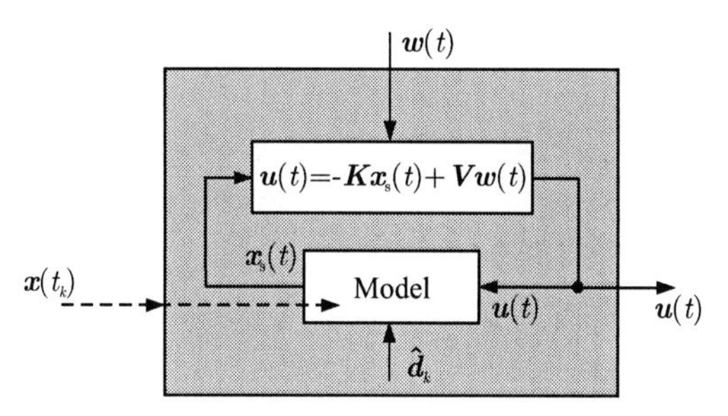

Figure 3.2.: Control input generator

The **control input generator** determines the input (3.11) by means of a model of the continuous-time closed-loop system (3.5)

$$\dot{\boldsymbol{x}}_{\mathrm{s}}(t) = \bar{\boldsymbol{A}}\boldsymbol{x}_{\mathrm{s}}(t) + \boldsymbol{E}\hat{\boldsymbol{d}}_k + \boldsymbol{B}\boldsymbol{V}\boldsymbol{w}(t), \qquad \boldsymbol{x}_{\mathrm{s}}(t_k^+) = \boldsymbol{x}(t_k),\ t \geq t_k \tag{3.12}$$

$$\boldsymbol{u}(t) = -\boldsymbol{K}\boldsymbol{x}_{\mathrm{s}}(t) + \boldsymbol{V}\boldsymbol{w}(t), \tag{3.13}$$

where $\boldsymbol{x}_{\mathrm{s}}$ is used throughout this work to denote *the state of the control input generator.* Henceforth, it is also called *model state* or *reference state.*

The time t_k^+ denotes the time instance immediately after the update of the model state $\boldsymbol{x}_{\mathrm{s}}$ with the measured state $\boldsymbol{x}(t_k)$, which the control input generator gets from the event generator at event time t_k. The control input $\boldsymbol{u}(t)$ is used as the output of the control input generator and, therefore, serves as the continuous input of the plant (3.1), (3.2). Figure 3.2 shows the block diagram of the control input generator. Suitable ways for determining the event time t_k and the disturbance estimate $\hat{\boldsymbol{d}}_k$ are presented in the Sections 3.3.3 and 3.3.4.

For discrete-time systems, a similar model-based approach has been investigated in [91, 92, 112].

3.3.2. Behaviour of the event-based state-feedback loop

The analysis in this section is valid for arbitrary event generators and arbitrary methods to estimate the disturbance magnitude $\hat{\boldsymbol{d}}_k$. It investigates the behaviour of the event-based control loop in the time interval $[t_k, t_{k+1})$ between the consecutive event times t_k and t_{k+1}.

The plant (3.1), (3.2) together with the control input generator (3.12), (3.13) is described for the time period $[t_k, t_{k+1})$ by the state-space model

$$\begin{aligned}
\begin{pmatrix} \dot{\boldsymbol{x}}(t) \\ \dot{\boldsymbol{x}}_{\mathrm{s}}(t) \end{pmatrix} &= \begin{pmatrix} \boldsymbol{A} & -\boldsymbol{BK} \\ \boldsymbol{O} & \bar{\boldsymbol{A}} \end{pmatrix} \begin{pmatrix} \boldsymbol{x}(t) \\ \boldsymbol{x}_{\mathrm{s}}(t) \end{pmatrix} + \begin{pmatrix} \boldsymbol{E} \\ \boldsymbol{O} \end{pmatrix} \boldsymbol{d}(t) + \begin{pmatrix} \boldsymbol{O} \\ \boldsymbol{E} \end{pmatrix} \hat{\boldsymbol{d}}_k \\
&\quad + \begin{pmatrix} \boldsymbol{BV} \\ \boldsymbol{BV} \end{pmatrix} \boldsymbol{w}(t), \qquad \begin{pmatrix} \boldsymbol{x}(t_k) \\ \boldsymbol{x}_{\mathrm{s}}(t_k^+) \end{pmatrix} = \begin{pmatrix} \boldsymbol{x}(t_k) \\ \boldsymbol{x}(t_k) \end{pmatrix} \\
\boldsymbol{y}(t) &= (\boldsymbol{C} \;\; \boldsymbol{O}) \begin{pmatrix} \boldsymbol{x}(t) \\ \boldsymbol{x}_{\mathrm{s}}(t) \end{pmatrix}.
\end{aligned}$$

This model takes into account that the closed-loop system is subject to the disturbance $\boldsymbol{d}(t)$, whereas the control input generator uses the disturbance estimate $\hat{\boldsymbol{d}}_k$, which is assumed to be constant in the time interval considered. The expression $\boldsymbol{x}(t_k) = \boldsymbol{x}(t_k)$ is used in the following to explicitly indicate that this state is not changed at the corresponding time instance.

To analyse the behaviour of this system, the state transformation

$$\begin{pmatrix} \boldsymbol{x}_\Delta(t) \\ \boldsymbol{x}_{\mathrm{s}}(t) \end{pmatrix} = \begin{pmatrix} \boldsymbol{I}_n & -\boldsymbol{I}_n \\ \boldsymbol{O} & \boldsymbol{I}_n \end{pmatrix} \begin{pmatrix} \boldsymbol{x}(t) \\ \boldsymbol{x}_{\mathrm{s}}(t) \end{pmatrix} \tag{3.14}$$

is used which introduces the difference state $\boldsymbol{x}_\Delta(t) = \boldsymbol{x}(t) - \boldsymbol{x}_{\mathrm{s}}(t)$. This matrix is regular and its inverse is given by

$$\begin{pmatrix} \boldsymbol{x}(t) \\ \boldsymbol{x}_{\mathrm{s}}(t) \end{pmatrix} = \begin{pmatrix} \boldsymbol{I}_n & \boldsymbol{I}_n \\ \boldsymbol{O} & \boldsymbol{I}_n \end{pmatrix} \begin{pmatrix} \boldsymbol{x}_\Delta(t) \\ \boldsymbol{x}_{\mathrm{s}}(t) \end{pmatrix}.$$

The transformed state-space model is

$$\begin{aligned}
\begin{pmatrix} \dot{\boldsymbol{x}}_\Delta(t) \\ \dot{\boldsymbol{x}}_\mathrm{s}(t) \end{pmatrix} &= \begin{pmatrix} \boldsymbol{A} & \boldsymbol{O} \\ \boldsymbol{O} & \bar{\boldsymbol{A}} \end{pmatrix} \begin{pmatrix} \boldsymbol{x}_\Delta(t) \\ \boldsymbol{x}_\mathrm{s}(t) \end{pmatrix} + \begin{pmatrix} \boldsymbol{E} \\ \boldsymbol{O} \end{pmatrix} \boldsymbol{d}(t) + \begin{pmatrix} -\boldsymbol{E} \\ \boldsymbol{E} \end{pmatrix} \hat{\boldsymbol{d}}_k \qquad (3.15) \\
&\quad + \begin{pmatrix} \boldsymbol{O} \\ \boldsymbol{BV} \end{pmatrix} \boldsymbol{w}(t), \quad \begin{pmatrix} \boldsymbol{x}_\Delta(t_k^+) \\ \boldsymbol{x}_\mathrm{s}(t_k^+) \end{pmatrix} = \begin{pmatrix} 0 \\ \boldsymbol{x}(t_k) \end{pmatrix} \\
\boldsymbol{y}(t) &= (\boldsymbol{C} \;\; \boldsymbol{C}) \begin{pmatrix} \boldsymbol{x}_\Delta(t) \\ \boldsymbol{x}_\mathrm{s}(t) \end{pmatrix}
\end{aligned}$$

whose output is given by the equation

$$\begin{aligned}
\boldsymbol{y}(t) &= \boldsymbol{C}\mathrm{e}^{\bar{\boldsymbol{A}}(t-t_k)} \boldsymbol{x}(t_k) + \int_{t_k}^{t} \boldsymbol{C}\mathrm{e}^{\boldsymbol{A}(t-\alpha)} \boldsymbol{E}(\boldsymbol{d}(\alpha) - \hat{\boldsymbol{d}}_k)\, \mathrm{d}\alpha \\
&\quad + \int_{t_k}^{t} \boldsymbol{C}\mathrm{e}^{\bar{\boldsymbol{A}}(t-\alpha)} \boldsymbol{E}\hat{\boldsymbol{d}}_k \,\mathrm{d}\alpha + \int_{t_k}^{t} \boldsymbol{C}\mathrm{e}^{\bar{\boldsymbol{A}}(t-\alpha)} \boldsymbol{BV}\boldsymbol{w}(\alpha)\,\mathrm{d}\alpha \\
&= \boldsymbol{C}\mathrm{e}^{\bar{\boldsymbol{A}}(t-t_k)} \boldsymbol{x}(t_k) + \int_{t_k}^{t} \boldsymbol{C}\mathrm{e}^{\boldsymbol{A}(t-\alpha)} \boldsymbol{E}\boldsymbol{d}_\Delta(\alpha)\,\mathrm{d}\alpha \qquad (3.16) \\
&\quad + \boldsymbol{C}\bar{\boldsymbol{A}}^{-1}\left(\mathrm{e}^{\bar{\boldsymbol{A}}(t-t_k)} - \boldsymbol{I}_n\right) \boldsymbol{E}\hat{\boldsymbol{d}}_k + \int_{t_k}^{t} \boldsymbol{C}\mathrm{e}^{\bar{\boldsymbol{A}}(t-\alpha)} \boldsymbol{BV}\boldsymbol{w}(\alpha)\,\mathrm{d}\alpha
\end{aligned}$$

with the disturbance estimation error

$$\boldsymbol{d}_\Delta(t) = \boldsymbol{d}(t) - \hat{\boldsymbol{d}}_k.$$

Accordingly, the output $\boldsymbol{y}(t)$ of the event-based control loop consists of two components

$$\boldsymbol{y}(t) = \boldsymbol{y}_\mathrm{s}(t) + \boldsymbol{y}_\Delta(t)$$

with

$$\begin{aligned}
\boldsymbol{y}_\mathrm{s}(t) &= \boldsymbol{C}\mathrm{e}^{\bar{\boldsymbol{A}}(t-t_k)} \boldsymbol{x}(t_k) + \boldsymbol{C}\bar{\boldsymbol{A}}^{-1}\left(\mathrm{e}^{\bar{\boldsymbol{A}}(t-t_k)} - \boldsymbol{I}_n\right) \boldsymbol{E}\hat{\boldsymbol{d}}_k \qquad (3.17) \\
&\quad + \int_{t_k}^{t} \boldsymbol{C}\mathrm{e}^{\bar{\boldsymbol{A}}(t-\alpha)} \boldsymbol{BV}\boldsymbol{w}(\alpha)\,\mathrm{d}\alpha \\
\boldsymbol{y}_\Delta(t) &= \int_{t_k}^{t} \boldsymbol{C}\mathrm{e}^{\boldsymbol{A}(t-\alpha)} \boldsymbol{E}\boldsymbol{d}_\Delta(\alpha)\,\mathrm{d}\alpha. \qquad (3.18)
\end{aligned}$$

Both components can be obtained from Eq. (3.15) together with the output equations

$$\boldsymbol{y}_{\mathrm{s}}(t) = (\boldsymbol{O}\ \boldsymbol{C})\begin{pmatrix}\boldsymbol{x}_\Delta(t)\\ \boldsymbol{x}_{\mathrm{s}}(t)\end{pmatrix}$$

$$\boldsymbol{y}_\Delta(t) = (\boldsymbol{C}\ \boldsymbol{O})\begin{pmatrix}\boldsymbol{x}_\Delta(t)\\ \boldsymbol{x}_{\mathrm{s}}(t)\end{pmatrix}.$$

Lemma 1. *The output of the event-based state-feedback loop* (3.1), (3.2), (3.12), (3.13) *subject to the disturbance* $\boldsymbol{d}(t) = \hat{\boldsymbol{d}}_k + \boldsymbol{d}_\Delta(t)$ *consists of two components* $\boldsymbol{y}(t) = \boldsymbol{y}_{\mathrm{s}}(t) + \boldsymbol{y}_\Delta(t)$ *with* $\boldsymbol{y}_{\mathrm{s}}(t)$ *given by Eq.* (3.17) *and* $\boldsymbol{y}_\Delta(t)$ *given by Eq.* (3.18).

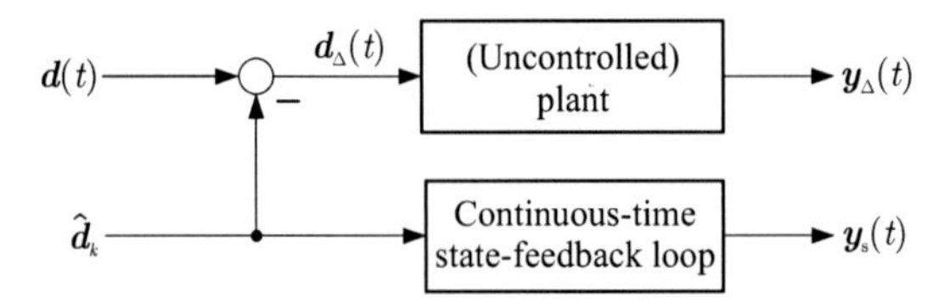

Figure 3.3.: Interpretation of Eq. (3.16)

This lemma shows three facts:

- The constant disturbance $\hat{\boldsymbol{d}}_k$, which is taken into account by the control input generator, has the same effect on the event-based control as on the continuous-time state feedback. In the time interval $[t_k, t_{k+1})$ with the initial state $\boldsymbol{x}(t_k)$ both systems generate the output $\boldsymbol{y}_{\mathrm{s}}(t)$.

- The difference $\boldsymbol{d}_\Delta(t) = \boldsymbol{d}(t) - \hat{\boldsymbol{d}}_k$ between the actual disturbance $\boldsymbol{d}(t)$ and the constant disturbance estimate $\hat{\boldsymbol{d}}_k$ affects the (uncontrolled) plant and results in the output $\boldsymbol{y}_\Delta(t)$, which describes the difference between the outputs of the continuous-time state-feedback loop and the event-based state-feedback loop (Fig. 3.3).

- For a good approximation $\hat{\boldsymbol{d}}_k$ of the disturbance $\boldsymbol{d}(t)$ in the time interval $[t_k, t_{k+1})$, i.e.

  $$\boldsymbol{d}(t) - \hat{\boldsymbol{d}}_k \approx \boldsymbol{0},$$

  the plant subject to the open-loop control (3.11) behaves like the continuous-time state-feedback loop. No communication is necessary in this time interval.

According to state transformation (3.14), the state trajectory $\boldsymbol{x}(t)$ of the event-based state-feedback loop can also be decomposed into two parts:

$$\boldsymbol{x}(t) = \boldsymbol{x}_{\mathrm{s}}(t) + \boldsymbol{x}_{\Delta}(t).$$

The state equation (3.15) yields

$$\begin{aligned} \boldsymbol{x}_{\mathrm{s}}(t) &= \mathrm{e}^{\bar{\boldsymbol{A}}(t-t_k)}\boldsymbol{x}(t_k) + \bar{\boldsymbol{A}}^{-1}\left(\mathrm{e}^{\bar{\boldsymbol{A}}(t-t_k)} - \boldsymbol{I}_n\right)\boldsymbol{E}\hat{\boldsymbol{d}}_k \\ &\quad + \int_{t_k}^{t} \mathrm{e}^{\bar{\boldsymbol{A}}(t-\alpha)}\boldsymbol{B}\boldsymbol{V}\boldsymbol{w}(\alpha)\,\mathrm{d}\alpha \qquad (3.19) \\ \boldsymbol{x}_{\Delta}(t) &= \int_{t_k}^{t} \mathrm{e}^{\boldsymbol{A}(t-\alpha)}\boldsymbol{E}\boldsymbol{d}_{\Delta}(\alpha)\,\mathrm{d}\alpha. \qquad (3.20) \end{aligned}$$

Note that $\boldsymbol{x}_{\mathrm{s}}(t)$ is identical to the state trajectory of the continuous-time state-feedback system (3.5), (3.6) in the time interval $[t_k, t_{k+1})$ with initial state $\boldsymbol{x}_{\mathrm{s}}(t_k) = \boldsymbol{x}(t_k)$ and affected by the constant disturbance $\boldsymbol{d}(t) = \hat{\boldsymbol{d}}_k$.

3.3.3. Event generation

This section presents the event generation which is applied in the event-based state-feedback loop. Events are generated by comparing the measured state trajectory $\boldsymbol{x}(t)$ with the state trajectory $\boldsymbol{x}_{\mathrm{s}}(t)$ that would occur in the continuous-time state-feedback loop for the constant disturbance $\boldsymbol{d}(t) = \hat{\boldsymbol{d}}_k$ and the feedback information $\boldsymbol{x}(t_k)$. As the state $\boldsymbol{x}_{\mathrm{s}}(t)$, which is determined by the control input generator according to Eq. (3.12), represents the desired reference signal, the measured state $\boldsymbol{x}(t)$ should be kept in a surrounding

$$\Omega_{\mathrm{s}}(\boldsymbol{x}_{\mathrm{s}}(t)) = \{\boldsymbol{x} : \|\boldsymbol{x} - \boldsymbol{x}_{\mathrm{s}}(t)\| \leq \bar{e}\}$$

of this state.

The **event generator** generates an event whenever the difference between the measured state $\boldsymbol{x}(t)$ and the reference state $\boldsymbol{x}_{\mathrm{s}}(t)$ reaches the event threshold $\bar{e}$:

$$\|\boldsymbol{x}(t) - \boldsymbol{x}_{\mathrm{s}}(t)\| = \bar{e}. \qquad (3.21)$$

At this time t, which is the event time t_k, the state information $\boldsymbol{x}(t_k)$ is communicated to the control input generator.

Note that $\boldsymbol{x}(t)$ is the measured state of the plant which is continuously available at the event generator. As the model state $\boldsymbol{x}_s(t)$ used in the control input generator is also required for the event generation, it has to be available in the event generator as well. In order not to continuously transmit the state information $\boldsymbol{x}_s(t)$ from the control input generator to the event generator, a copy of the control input generator is included in the event generator and $\boldsymbol{x}_s(t)$ is also determined by the event generator by means of Eq. (3.12). This is illustrated in Fig. 3.4.

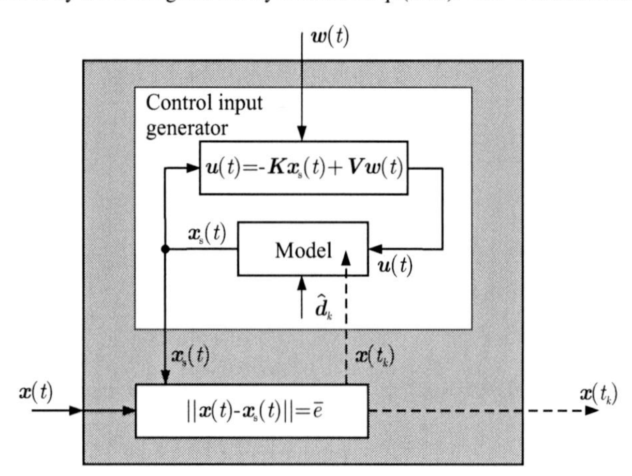

Figure 3.4.: Event generator

The event condition proposed leads to the following property.

Lemma 2. *Event condition* (3.21) *ensures that the difference state* $\boldsymbol{x}_\Delta(t) = \boldsymbol{x}(t) - \boldsymbol{x}_s(t)$ *is bounded and remains for all times* $t \geq 0$ *in the set* Ω_Δ*:*

$$\boldsymbol{x}_\Delta(t) \in \Omega_\Delta = \{\boldsymbol{x}_\Delta \, : \, \|\boldsymbol{x}_\Delta\| \leq \bar{e}\}, \quad \forall t.$$

Proof. Whenever the difference state $\boldsymbol{x}_\Delta(t)$ touches the boundary

$$\partial\Omega_\Delta = \{\boldsymbol{x}_\Delta \, : \, \|\boldsymbol{x}_\Delta\| = \bar{e}\}$$

of the set Ω_Δ, the communication mechanism resets the model state $\boldsymbol{x}_s(t_k)$ according to $\boldsymbol{x}_s(t_k^+) = \boldsymbol{x}(t_k)$ which yields $\boldsymbol{x}_\Delta(t_k^+) = \mathbf{0}$. Hence, $\|\boldsymbol{x}_\Delta(t)\| \leq \bar{e}$ holds for all times $t \geq 0$. □

3.3.4. Disturbance estimation

The following investigations show how to get an estimate $\hat{\boldsymbol{d}}_k$ of the disturbance magnitude at the event time t_k. In the preceding time interval $[t_{k-1}, t_k)$ the disturbance estimate $\hat{\boldsymbol{d}}_{k-1}$ has been used. Consider now the difference $\boldsymbol{x}_\Delta(t) = \boldsymbol{x}(t) - \boldsymbol{x}_{\rm s}(t)$ and assume that the disturbance $\boldsymbol{d}(t)$ has been constant in this time interval

$$\boldsymbol{d}(t) = \bar{\boldsymbol{d}}, \quad \text{for } t \in [t_{k-1}, t_k)$$

where $\bar{\boldsymbol{d}}$ is the actual disturbance magnitude, which usually differs from the estimate $\hat{\boldsymbol{d}}_k$. Equation (3.20) yields

$$\begin{aligned} \boldsymbol{x}(t) - \boldsymbol{x}_{\rm s}(t) &= \int_{t_{k-1}}^{t} {\rm e}^{\boldsymbol{A}(t-\alpha)} \boldsymbol{E}(\bar{\boldsymbol{d}} - \hat{\boldsymbol{d}}_{k-1})\, {\rm d}\alpha \\ &= \boldsymbol{A}^{-1}\left({\rm e}^{\boldsymbol{A}(t-t_{k-1})} - \boldsymbol{I}_n\right) \boldsymbol{E}(\bar{\boldsymbol{d}} - \hat{\boldsymbol{d}}_{k-1}). \end{aligned}$$

This equation is used to determine, at time $t = t_k$, the unknown disturbance magnitude $\bar{\boldsymbol{d}}$ according to

$$\bar{\boldsymbol{d}} = \hat{\boldsymbol{d}}_{k-1} + \left(\boldsymbol{A}^{-1}\left({\rm e}^{\boldsymbol{A}(t_k-t_{k-1})} - \boldsymbol{I}_n\right) \boldsymbol{E}\right)^{+} \left(\boldsymbol{x}(t_k) - \boldsymbol{x}_{\rm s}(t_k)\right). \tag{3.22}$$

The **disturbance estimator** uses the disturbance magnitude $\bar{\boldsymbol{d}}$ derived by Eq. (3.22) as the estimate $\hat{\boldsymbol{d}}_k$ in the time interval $t \geq t_k$:

$$\hat{\boldsymbol{d}}_0 = 0 \tag{3.23}$$

$$\hat{\boldsymbol{d}}_k = \hat{\boldsymbol{d}}_{k-1} + \left(\boldsymbol{A}^{-1}\left({\rm e}^{\boldsymbol{A}(t_k-t_{k-1})} - \boldsymbol{I}_n\right) \boldsymbol{E}\right)^{+} \left(\boldsymbol{x}(t_k) - \boldsymbol{x}_{\rm s}(t_k)\right). \tag{3.24}$$

As initially no information about the disturbance is available, $\hat{\boldsymbol{d}}_0$ is chosen to be zero.

The pseudoinverse in Eq. (3.24) exists if, as usual, the number of disturbances is lower than the number of state variables $(n \geq l)$ and the occurring matrices have full rank. Note that this disturbance estimation explicitly requires the existence of the inverse system matrix $\boldsymbol{A}^{-1}$, which will be relaxed in Section 4.4.2 by introducing an alternative disturbance observer.

The disturbance estimator is included in the control input generator as well as in the event generator to provide both components with a current disturbance estimate at event times t_k

$(k = 0, 1, 2, ..)$, which is depicted in Fig. 3.5.

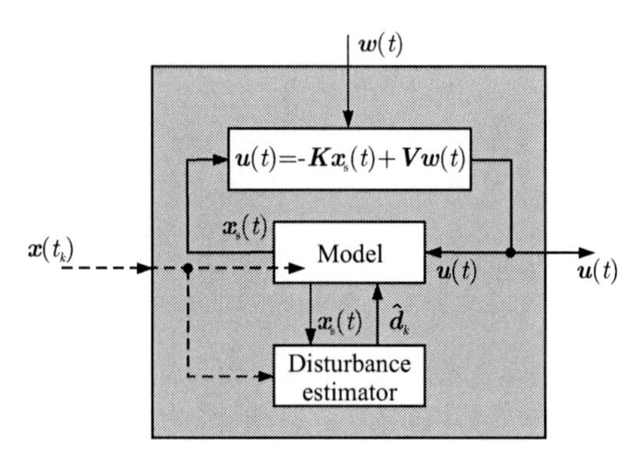

Figure 3.5.: Control input generator with disturbance estimator

Properties of the disturbance estimator. The following result states that for a constant disturbance

$$\boldsymbol{d}(t) = \bar{\boldsymbol{d}}, \quad \forall t \tag{3.25}$$

the disturbance magnitude $\bar{\boldsymbol{d}}$ is correctly determined by the disturbance estimator (3.23), (3.24) once the first event has been alerted.

Lemma 3. *If at time t_1 the first event has been generated in the event-based state-feedback loop affected by a constant disturbance* (3.25)*, the disturbance estimator* (3.23), (3.24) *correctly determines the disturbance magnitude:*

$$\hat{\boldsymbol{d}}_1 = \bar{\boldsymbol{d}}. \tag{3.26}$$

Proof. See Appendix B.1, page 171. □

For a scalar time-varying disturbance $d(t)$ the disturbance estimation is exemplarily illustrated in Fig. 3.6 which shows the behaviour of the disturbance $d(t)$ and the corresponding sequence of disturbance estimates. Here, $\hat{d}_1$ is the weighted average of the disturbance $d(t)$ for the time interval $[t_0, t_1)$. Similarly, $\hat{d}_2$ describes a weighted average of the actual disturbance for the time interval $[t_1, t_2)$. If the disturbance remains constant over two time intervals, then in the second time interval the estimate $\hat{d}_k$ coincides with the true magnitude of the disturbance. This happens in the example for $t \geq t_4$.

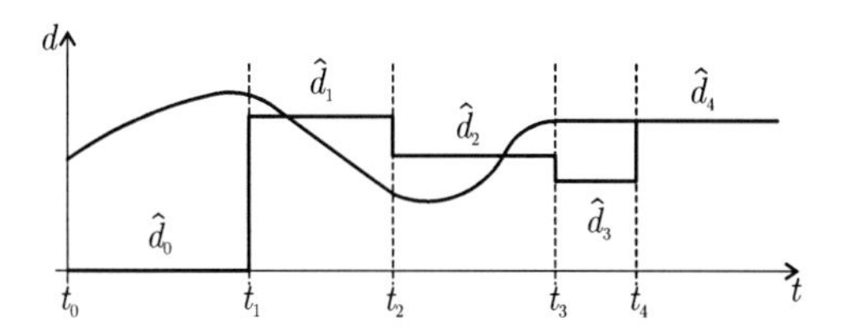

Figure 3.6.: Disturbance $d(t)$ and disturbance estimates $\hat{d}_0, \hat{d}_1, \hat{d}_2$...

3.3.5. Summary of the components of the event-based state-feedback loop

The event-based state-feedback loop as considered in this thesis uses the structure depicted in Fig. 3.1 which consists of

- the plant (3.1), (3.2),
- the control input generator (3.12), (3.13) which also estimates the disturbance according to Eqs. (3.23), (3.24), and
- the event generator which includes a copy of the control input generator (3.12), (3.13) and the disturbance estimator (3.23), (3.24) and determines the event times t_k according to Eq. (3.21).

At event times t_k $(k = 0, 1, 2, ...)$ the measured state information $\boldsymbol{x}(t_k)$ is sent from the event generator towards the control input generator and is used there as well as in the event generator to update the model state $\boldsymbol{x}_s$ according to $\boldsymbol{x}_s(t_k^+) = \boldsymbol{x}(t_k)$ and to determine the new disturbance estimate $\hat{\boldsymbol{d}}_k$. Since it is assumed that there is no time delay between the event generation and the receipt of the data $\boldsymbol{x}(t_k)$ by the control input generator, the models in the control input generator and the event generator work synchronously for all times t.

Example 1 *Behaviour of the event-based state-feedback loop*

Figure 3.7 presents simulation results which illustrate the behaviour of the event-based state-feedback loop in different scenarios. The plant is given by the model of the thermofluid process (2.7), (2.8) with $\boldsymbol{x}_0 = \mathbf{0}$ and the controller is chosen to be

$$\boldsymbol{K} = \begin{pmatrix} 0.08 & -0.02 \\ 0.17 & 0.72 \end{pmatrix}, \tag{3.27}$$

for which the continuous-time closed-loop system (3.5), (3.6) is stable and has desirable disturbance attenuation properties. The event threshold $\bar{e}$ is set to $\bar{e} = 2$ and the event generator uses the supremum norm, i.e. the event condition is given by

$$\|\boldsymbol{x}_\Delta(t)\|_\infty = \|\boldsymbol{x}(t) - \boldsymbol{x}_\mathrm{s}(t)\|_\infty = 2. \tag{3.28}$$

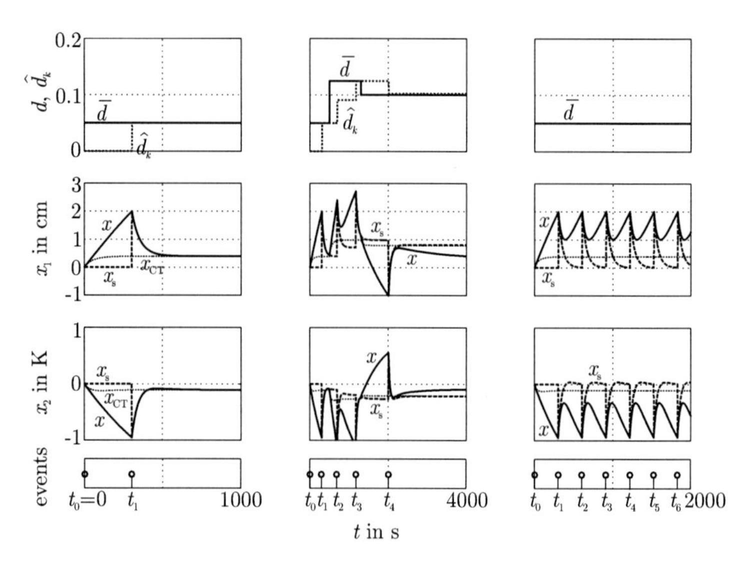

Figure 3.7.: Behaviour of the event-based state-feedback loop subject to a constant disturbance (left plots), a time-varying disturbance (middle plots) and without applying the disturbance estimation (right plots). Solid lines: plant state $\boldsymbol{x}(t)$; dashed lines: model state $\boldsymbol{x}_\mathrm{s}(t)$; dotted lines: state $\boldsymbol{x}_\mathrm{CT}(t)$ of the continuous-time state-feedback loop.

In the first investigation (left-hand side of Fig. 3.7) the plant is subject to a constant disturbance $d(t) = \bar{d}$ drawn by the solid line in the top subplot. After the initialising event at time $t_0 = 0$, an event takes place at time t_1 due to the level behaviour, where

$$|x_{\Delta,1}(t_1)| = |x_1(t_1) - x_{\mathrm{s},1}(t_1)| = 2 \tag{3.29}$$

holds (see second subplot from top). At this event time, which is indicated in the bottom subplot, the disturbance magnitude $\bar{d}$ is correctly estimated by the disturbance estimator ($\hat{d}_1 = \bar{d}$, dotted line) and the state information $\boldsymbol{x}(t_1)$ is communicated to the control input generator. Afterwards, both $\boldsymbol{x}(t)$ and $\boldsymbol{x}_\mathrm{s}(t)$ coincide and behave stationarily like the state $\boldsymbol{x}_\mathrm{CT}(t)$ of the continuous-time state-feedback loop. No further event occurs. The steady-state control error occurs due to the fact that a proportional state feedback is used. It can be avoided by using controllers with integral action which is investigated in Section 4.3.

The third subplot shows the temperature behaviour. Note that the minus sign results from the fact that a deviation from the operating point (2.6) is considered. This likewise holds for the level behaviour and, moreover, also in the following examples.

In the second investigation (middle plots of Fig. 3.7) the disturbance magnitude changes after the first events as shown by the solid line in the top subplot. Five events take place until the disturbance remains constant and its magnitude is estimated with sufficient accuracy.

Again, the events are generated because equality (3.29) is satisfied at the event times t_k ($k = 1, 2, 3, 4$). At time t_1, the estimate $\hat{d}_1$ of the disturbance $d(t)$ in the preceding time interval $[0, t_1)$ is determined by the disturbance estimator. As the disturbance varies, the estimate $\hat{d}_1$ and the disturbance $d(t)$ differ for $t \geq t_1$ and a new event occurs at time t_2. The new disturbance estimate $\hat{d}_2$ describes a weighted average of $d(t)$ in the preceding time interval $[t_1, t_2)$, which also holds at the subsequent event times (Section 3.3.4).

The right plots of Fig. 3.7 show the importance of the disturbance estimation (3.23), (3.24) included in the control input generation and the event generation. Here, the disturbance $d(t)$ is considered to be constant with the same magnitude as in the left plots but the control input generator does not apply the disturbance estimation according to Eq. (3.24). Instead, it uses the disturbance estimate $\hat{d}_0 = 0$ for all times t.

As a consequence, after the first event at time t_1 further events occur because the control input generator provides an input $\boldsymbol{u}(t)$ which is generated without taking the current disturbance into account. Therefore, the plant state $\boldsymbol{x}(t)$ and the model state $\boldsymbol{x}_\mathrm{s}(t)$ diverge and events are generated periodically for all times $t > t_1$ because the disturbance $d(t)$ remains constant.

3.4. Main properties of the event-based state feedback

The central properties to be investigated when considering event-based control concern the stability and the communication over the feedback link. Henceforth, both properties are investigated for the event-based state feedback based on the following two methods:

- The stability properties are investigated by comparing the behaviour of the event-based state-feedback loop with the desired behaviour of the continuous-time state-feedback loop.
- The communication properties of the event-based state-feedback loop are investigated by analysing the behaviour of the difference state $\boldsymbol{x}_\Delta(t)$ of the transformed model (3.15) together with the event condition (3.21).

The main results of the subsequent analysis can be summarised as follows:

- The state $\boldsymbol{x}(t)$ of the event-based control loop is *GUUB* and there exists an upper bound on the approximation error (Theorem 3, Theorem 4).
- There exists a lower bound on the minimum inter-event time (Theorem 5).
- There exists a disturbance magnitude for which the event generator does not generate any event (Lemma 5).
- Setpoint tracking can be guaranteed only under specific conditions (Theorem 6).
- The event-based state-feedback loop is robust in the sense that its state $\boldsymbol{x}(t)$ remains *GUUB* for certain model uncertainties (Theorem 7).

3.4.1. Comparison of the event-based state-feedback loop and the continuous-time state-feedback loop

The quality of the event-based state feedback is evaluated in this section by showing that the deviation of the event-based control loop from the behaviour of the continuous-time state-feedback loop is bounded and can be made arbitrarily small by appropriately choosing the threshold $\bar{e}$ of the event generator (3.21). The continuous-time state-feedback loop is described by Eqs. (3.5), (3.6) and its state is denoted by $\boldsymbol{x}_{\mathrm{CT}}(t)$. The event-based control loop is described by Eqs. (3.1), (3.2) and Eq. (3.13), which lead to the state equation

$$\dot{\boldsymbol{x}}(t) = \boldsymbol{A}\boldsymbol{x}(t) - \boldsymbol{B}\boldsymbol{K}\boldsymbol{x}_{\mathrm{s}}(t) + \boldsymbol{E}\boldsymbol{d}(t) + \boldsymbol{B}\boldsymbol{V}\boldsymbol{w}(t), \qquad \boldsymbol{x}(0) = \boldsymbol{x}_0. \tag{3.30}$$

The evaluation is made by first showing that the approximation error

$$\boldsymbol{e}(t) = \boldsymbol{x}(t) - \boldsymbol{x}_{\mathrm{CT}}(t)$$

between the state $\boldsymbol{x}(t)$ of the event-based control loop and the state $\boldsymbol{x}_{\mathrm{CT}}(t)$ of the continuous-time control loop is bounded and second by deriving an upper bound of $\|\boldsymbol{e}(t)\|$.

For the state difference $\boldsymbol{e}(t)$ the equations

$$\begin{aligned}
\dot{\boldsymbol{e}}(t) &= \dot{\boldsymbol{x}}(t) - \dot{\boldsymbol{x}}_{\mathrm{CT}}(t) \\
&= \boldsymbol{A}\boldsymbol{x}(t) - \boldsymbol{B}\boldsymbol{K}\boldsymbol{x}_{\mathrm{s}}(t) + \boldsymbol{E}\boldsymbol{d}(t) + \boldsymbol{B}\boldsymbol{V}\boldsymbol{w}(t) - \bar{\boldsymbol{A}}\boldsymbol{x}_{\mathrm{CT}}(t) - \boldsymbol{E}\boldsymbol{d}(t) - \boldsymbol{B}\boldsymbol{V}\boldsymbol{w}(t) \\
&= \boldsymbol{A}(\boldsymbol{x}(t) - \boldsymbol{x}_{\mathrm{CT}}(t)) - \boldsymbol{B}\boldsymbol{K}(\boldsymbol{x}_{\mathrm{s}}(t) - \boldsymbol{x}_{\mathrm{CT}}(t)) \\
&= \boldsymbol{A}\boldsymbol{e}(t) - \boldsymbol{B}\boldsymbol{K}(\boldsymbol{x}(t) - \boldsymbol{x}_{\Delta}(t) - \boldsymbol{x}_{\mathrm{CT}}(t)) \\
&= \bar{\boldsymbol{A}}\boldsymbol{e}(t) + \boldsymbol{B}\boldsymbol{K}\boldsymbol{x}_{\Delta}(t), \quad \boldsymbol{e}(0) = \boldsymbol{0} \qquad (3.31)
\end{aligned}$$

hold, where $\boldsymbol{x}_{\mathrm{s}}(t)$ was replaced by $\boldsymbol{x}_{\mathrm{s}}(t) = \boldsymbol{x}(t) - \boldsymbol{x}_{\Delta}(t)$ according to state transformation (3.14).

As the state $\boldsymbol{x}_{\Delta}(t)$ is bounded according to Lemma 2 and the matrix $\bar{\boldsymbol{A}}$ is Hurwitz, the difference $\boldsymbol{e}(t)$ is bounded according to Theorem 2.

Lemma 4. *The approximation error $\boldsymbol{e}(t)$ between the state $\boldsymbol{x}(t)$ of the event-based state-feedback loop and the state $\boldsymbol{x}_{\mathrm{CT}}(t)$ of the continuous-time state-feedback loop is bounded.*

Note that this result holds independently of any kind of disturbances or command inputs that affect the plant. The analysis can be made more precise. As, according to Lemma 2, the state $\boldsymbol{x}_{\Delta}(t)$ is bounded by $\bar{e}$, the bound on the difference $\boldsymbol{e}(t)$ depends monotonically on $\bar{e}$. To see this, the difference $\boldsymbol{e}(t)$ is represented by the convolution

$$\boldsymbol{e}(t) = \int_0^t \mathrm{e}^{\bar{\boldsymbol{A}}(t-\alpha)} \boldsymbol{B}\boldsymbol{K}\boldsymbol{x}_{\Delta}(\alpha)\,\mathrm{d}\alpha,$$

of the model (3.31) [79]. Thus, the inequalities

$$\begin{aligned}
\|\boldsymbol{e}(t)\| &\leq \int_0^t \left\|\mathrm{e}^{\bar{\boldsymbol{A}}(t-\alpha)} \boldsymbol{B}\boldsymbol{K}\right\| \cdot \|\boldsymbol{x}_{\Delta}(\alpha)\|\,\mathrm{d}\alpha \\
&\leq \bar{e} \cdot \int_0^\infty \left\|\mathrm{e}^{\bar{\boldsymbol{A}}\alpha} \boldsymbol{B}\boldsymbol{K}\right\|\,\mathrm{d}\alpha = e_{\max} \qquad (3.32)
\end{aligned}$$

hold and the following theorem can be obtained.

Theorem 3. *The approximation error* $\mathbf{e}(t) = \boldsymbol{x}(t) - \boldsymbol{x}_{\mathrm{CT}}(t)$ *between the state* $\boldsymbol{x}(t)$ *of the event-based state-feedback loop* (3.1), (3.2), (3.12), (3.13), (3.21), (3.23), (3.24) *and the state* $\boldsymbol{x}_{\mathrm{CT}}(t)$ *of the continuous-time state-feedback loop* (3.5), (3.6) *is bounded from above by*

$$\|\boldsymbol{e}(t)\| \leq e_{\max}$$

with $e_{\max}$ *given by Eq.* (3.32).

This theorem shows that the event-based state feedback can be made to mimic a continuous-time state feedback with arbitrary precision by accordingly choosing the event threshold $\bar{e}$. The state $\boldsymbol{x}(t)$ remains in the bounded surrounding

$$\boldsymbol{x}(t) \in \Omega_{\mathrm{e}}(\boldsymbol{x}_{\mathrm{CT}}(t)) = \{\boldsymbol{x} : \|\boldsymbol{x} - \boldsymbol{x}_{\mathrm{CT}}(t)\| \leq e_{\max}\}$$

of the state $\boldsymbol{x}_{\mathrm{CT}}(t)$ of the continuous-time state-feedback loop for all times t, which is depicted in Fig. 3.8.

Hence, as the state $\boldsymbol{x}_{\mathrm{CT}}(t)$ of the continuous-time state-feedback loop is *GUUB* (Section 3.2.1), the state $\boldsymbol{x}(t)$ of the event-based state-feedback loop is *GUUB* as well.

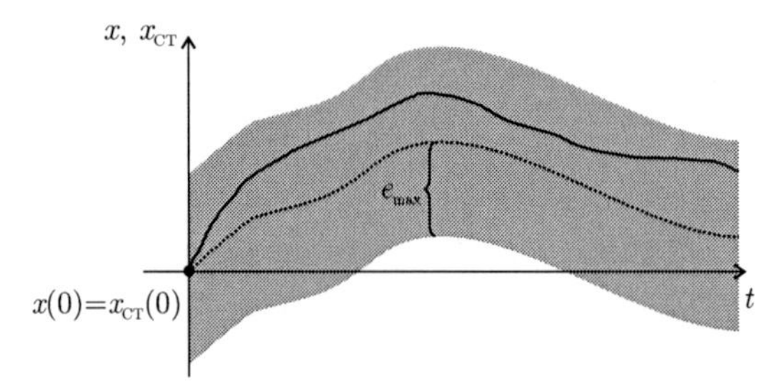

Figure 3.8.: Illustration of the error bound $e_{\max}$. Solid line: state $\boldsymbol{x}(t)$ of the event-based control loop; dotted line: state $\boldsymbol{x}_{\mathrm{CT}}(t)$ of the continuous-time control loop.

Equation (3.32) can be used to determine for every tolerable upper bound on the approximation error $\|\boldsymbol{e}(t)\|$ the event threshold $\bar{e}$. The price for a higher precision (smaller $e_{\max}$) is a more frequent communication between the event generator and the control input generator. This can be expected for every event-based control scheme, which is illustrated in the following example and investigated for the event-based state feedback in Section 3.4.3.

Example 2 *Approximation accuracy of the event-based state feedback*

For the thermofluid process (2.7), (2.8) with $\boldsymbol{x}_0 = \mathbf{0}$, controller (3.27) and the event condition (3.28) with the event threshold $\bar{e} = 2$, the upper error bound

$$\|\boldsymbol{e}(t)\|_\infty \leq e_{\max,2} = 2.26$$

results according to Eq. (3.32).

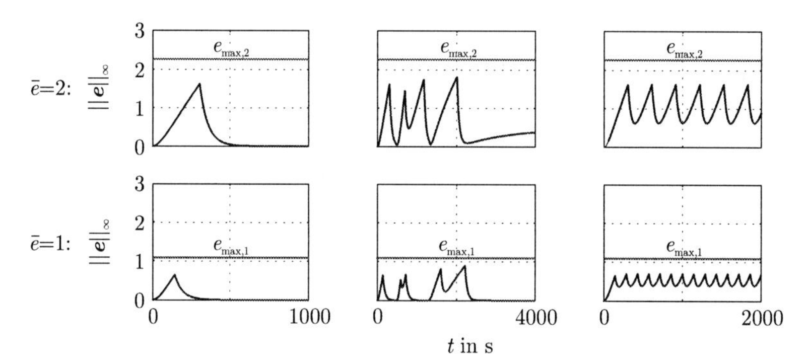

Figure 3.9.: Approximation error $\|\boldsymbol{e}(t)\|_\infty$ of the event-based state-feedback loop subject to a constant disturbance (left plots), a time-varying disturbance (middle plots) and without applying the disturbance estimation (right plots) for $\bar{e} = 2$ and $\bar{e} = 1$.

The actual approximation accuracy of the event-based state feedback in the situations considered in Example 1 is shown in the upper plots of Fig. 3.9. Note that every peak in these subplots corresponds to the respective event time in Fig. 3.7. It can be seen that the error $\boldsymbol{e}(t)$ increases before an event occurs, immediately decreases afterwards, and never reaches the error bound $e_{\max,2}$.

To outline the effect of adjusting the event threshold, the lower plots show the approximation error for $\bar{e} = 1$ which yields the error bound

$$e_{\max,1} = 1.13.$$

As expected, the actual approximation error decreases and never reaches $e_{\max,1}$. However, the communication increases which can be seen in all three cases as the first event at time t_1 occurs always earlier than for $\bar{e} = 2$. Using no disturbance estimation (right subplots), this becomes even more evident as instead of seven events overall 14 events occur.

3.4.2. Lyapunov-based stability analysis

The stability analysis in the previous section was based on a comparison between the behaviour of the event-based state-feedback loop and the continuous-time state-feedback loop. This section presents an alternative way which uses Lyapunov functions to show that the state $\boldsymbol{x}(t)$ of the event-based control loop is *GUUB*.

The starting point of the following analysis is given by the state equation (3.30) of the event-based control loop. This equation can be rewritten by replacing $\boldsymbol{x}_{\mathrm{s}}(t)$ according to $\boldsymbol{x}_{\mathrm{s}}(t) = \boldsymbol{x}(t) - \boldsymbol{x}_{\Delta}(t)$ which yields

$$\dot{\boldsymbol{x}}(t) = \bar{\boldsymbol{A}}\boldsymbol{x}(t) + \boldsymbol{B}\boldsymbol{K}\boldsymbol{x}_{\Delta}(t) + \boldsymbol{E}\boldsymbol{d}(t) + \boldsymbol{B}\boldsymbol{V}\boldsymbol{w}(t), \qquad \boldsymbol{x}(0) = \boldsymbol{x}_0. \tag{3.33}$$

Since $\bar{\boldsymbol{A}}$ is assumed to be Hurwitz, there always exists, for any given positive definite matrix $\boldsymbol{Q}$, a positive definite matrix $\boldsymbol{P}$ which satisfies the Lyapunov equation [80]

$$\bar{\boldsymbol{A}}'\boldsymbol{P} + \boldsymbol{P}\bar{\boldsymbol{A}} = -\boldsymbol{Q}.$$

Consider the quadratic Lyapunov function candidate

$$V(\boldsymbol{x}(t)) = \boldsymbol{x}'(t)\boldsymbol{P}\boldsymbol{x}(t)$$

whose time derivative is given by

$$\begin{aligned}
\dot{V}(\boldsymbol{x}(t)) &= (\boldsymbol{x}'(t)\bar{\boldsymbol{A}}' + \boldsymbol{x}_{\Delta}{}'(t)\boldsymbol{K}'\boldsymbol{B}' + \boldsymbol{d}'(t)\boldsymbol{E}' + \boldsymbol{w}'(t)\boldsymbol{V}'\boldsymbol{B}')\boldsymbol{P}\boldsymbol{x}(t)\\
&\quad + \boldsymbol{x}'(t)\boldsymbol{P}(\bar{\boldsymbol{A}}\boldsymbol{x}(t) + \boldsymbol{B}\boldsymbol{K}\boldsymbol{x}_{\Delta}(t) + \boldsymbol{E}\boldsymbol{d}(t) + \boldsymbol{B}\boldsymbol{V}\boldsymbol{w}(t))\\
&= -\boldsymbol{x}'(t)\boldsymbol{Q}\boldsymbol{x}(t) + 2\boldsymbol{x}'(t)\boldsymbol{P}\boldsymbol{B}\boldsymbol{K}\boldsymbol{x}_{\Delta}(t) + 2\boldsymbol{x}'(t)\boldsymbol{P}\boldsymbol{E}\boldsymbol{d}(t) + 2\boldsymbol{x}'(t)\boldsymbol{P}\boldsymbol{B}\boldsymbol{V}\boldsymbol{w}(t)\\
&\leq -c_{\mathrm{Q}}||\boldsymbol{x}(t)||^2 + c_{\mathrm{P}}||\boldsymbol{x}(t)|| \cdot (\delta_{\bar{e}} + \delta_{\mathrm{d}} + \delta_{\mathrm{w}})
\end{aligned}$$

with

$$\begin{aligned}
c_{\mathrm{Q}} &= \lambda_{\min}(\boldsymbol{Q})\\
c_{\mathrm{P}} &= \lambda_{\max}(\boldsymbol{P})\\
\delta_{\bar{e}} &= 2||\boldsymbol{B}\boldsymbol{K}||\bar{e}\\
\delta_{\mathrm{d}} &= 2||\boldsymbol{E}||d_{\max}\\
\delta_{\mathrm{w}} &= 2||\boldsymbol{B}\boldsymbol{V}||w_{\max}.
\end{aligned}$$

Here, $||\boldsymbol{x}_{\Delta}(t)||$ was replaced by $\bar{e}$ according to Lemma 2 and $w_{\max}$ denotes the maximum bound for the admissible command input $\boldsymbol{w}(t)$:

$$\|\boldsymbol{w}(t)\| \leq w_{\max}.$$

It follows that for all states $\boldsymbol{x}(t)$ which satisfy

$$||\boldsymbol{x}(t)|| > \frac{c_{\mathrm{P}}}{c_{\mathrm{Q}}}(\delta_{\bar{e}} + \delta_{\mathrm{d}} + \delta_{\mathrm{w}}) = p_{\mathrm{L}} \tag{3.34}$$

the relation $\dot{V}(\boldsymbol{x}(t)) < 0$ is obtained. Therefore, it exists a time $\bar{t}$ such that

$$\boldsymbol{x}(t) \in \Omega_{\mathrm{t}} = \{\boldsymbol{x} \,:\, \|\boldsymbol{x}\| \leq p_{\mathrm{L}}\} \quad \forall t \geq \bar{t} \tag{3.35}$$

holds, which leads to the main result of this section.

Theorem 4. *The state $\boldsymbol{x}(t)$ of the event-based state-feedback loop* (3.1), (3.2), (3.12), (3.13), (3.21), (3.23), (3.24) *is GUUB and remains for all times $t \geq \bar{t}$ in the set Ω_{t} given by Eqs.* (3.34), (3.35).

Compared to the stability result obtained in the previous section, the main benefit of this result is that it allows to directly determine the target set Ω_{t} (see Definition 2, page 19) for the event-based control loop without the requirement of firstly deriving the target set for the continuous-time closed-loop system.

Remark. The Lyapunov approach can be analogously applied to the difference behaviour $\boldsymbol{e}(t)$ given by Eq. (3.31) in order to specify the error bound $e_{\max}$ (Theorem 3). As both the disturbance $\boldsymbol{d}(t)$ and the command input $\boldsymbol{w}(t)$ do not appear in Eq. (3.31), $e_{\max}$ can be determined by using relation (3.34) and cancelling δ_{d} and δ_{w}.

Proposition 1. *As $\boldsymbol{e}(0) = \mathbf{0}$ holds, the approximation error $\boldsymbol{e}(t) = \boldsymbol{x}(t) - \boldsymbol{x}_{\mathrm{CT}}(t)$ between the state $\boldsymbol{x}(t)$ of the event-based state-feedback loop and the state $\boldsymbol{x}_{\mathrm{CT}}(t)$ of the continuous-time state-feedback loop* (3.5), (3.6) *is bounded from above by*

$$\|\boldsymbol{e}(t)\| \leq e_{\max,\mathrm{L}} = \frac{c_{\mathrm{P}}}{c_{\mathrm{Q}}}\delta_{\bar{e}}, \quad \forall t. \tag{3.36}$$

Vice versa, Eq. (3.33) can be also used together with the analysis in Section 3.4.1 to show that the state $\boldsymbol{x}(t)$ of the event-based control loop is *GUUB* and to directly derive the target set

Ω_t.

Proposition 2. *The state $\boldsymbol{x}(t)$ of the event-based state-feedback loop is GUUB and remains, for all times $t \geq \bar{t}$, in the set*

$$\boldsymbol{x}(t) \in \Omega_t = \{\boldsymbol{x} : \|\boldsymbol{x}\| \leq p_e\}$$

with

$$p_e = (\|\boldsymbol{BK}\|\bar{e} + \|\boldsymbol{E}\|d_{\max} + \|\boldsymbol{BV}\|w_{\max}) \cdot \int_0^\infty \|e^{\bar{\boldsymbol{A}}\alpha}\| \, d\alpha. \tag{3.37}$$

Example 3 *Determination of the target set Ω_t*

Consider the thermofluid process (2.7), (2.8) with the nonzero initial condition

$$\boldsymbol{x}_0 = \begin{pmatrix} 8 \\ 7 \end{pmatrix},$$

controller (3.27) and the event condition (3.28) with the event threshold $\bar{e} = 2$. With

$$\frac{c_P}{c_Q} = 37.78,$$

$d_{\max} = 0.15$ and $w_{\max} = 0$, Eq. (3.34) yields

$$p_L = 5.$$

The behaviour of the event-based state-feedback loop subject to a time-varying disturbance $\boldsymbol{d}(t)$ which satisfies $\|\boldsymbol{d}(t)\| \leq d_{\max}$ is depicted in the state space in Fig. 3.10. It shows that the state $\boldsymbol{x}(t)$ is GUUB and remains stationarily in the specified target set Ω_t. In this case, the target set is quadratic because the supremum norm has been used to derive p_L.

Interestingly, Eq. (3.37) yields almost the same parameter:

$$p_e \approx p_L = 5.$$

However, usually the Lyapunov approach may be more conservative which can be seen by deriving the error bound $e_{\max}$ by means of Eqs. (3.32), (3.36). It holds

$$e_{\max} = 2.26 < e_{\max,L} = 3.3$$

(cf. Example 2).

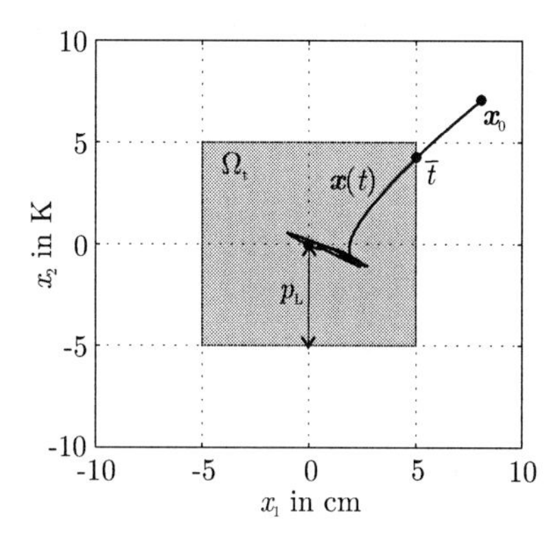

Figure 3.10.: Behaviour of the event-based state-feedback loop in the state space

3.4.3. Minimum inter-event time

The information exchange over the feedback link in the event-based control loop is the second main property to be investigated when considering event-based control. Basically, it has to be avoided that the inter-event times become arbitrarily small or that the event-based control loop exhibits Zeno behaviour [85], i.e. events are generated infinitely often in a finite time interval.

This section derives the result that the minimum inter-event time of the event-based state-feedback loop is bounded from below and depends upon the disturbance $\boldsymbol{d}(t)$. For a bounded disturbance, the minimum inter-event time can be evaluated as follows. The difference state $\boldsymbol{x}_\Delta(t)$ (Eq. (3.15)) is described by the state-space model

$$\dot{\boldsymbol{x}}_\Delta(t) = \boldsymbol{A}\boldsymbol{x}_\Delta(t) + \boldsymbol{E}\boldsymbol{d}_\Delta(t), \quad \boldsymbol{x}_\Delta(t_k^+) = \boldsymbol{0}, \;\; t \in [t_k, t_{k+1}).$$

Assume that the disturbance estimation error $\boldsymbol{d}_\Delta(t) = \boldsymbol{d}(t) - \hat{\boldsymbol{d}}_k$ satisfies

$$\|\boldsymbol{d}_\Delta(t)\| \leq \gamma\, d_{\max} \quad \text{for} \;\; t \geq 0, \tag{3.38}$$

where γ is a parameter which depends on the possible rate of change of the disturbance between two consecutive events. Hence,

$$0 \leq \gamma \leq 2$$

holds since $\gamma\, d_{\max}$ is a bound on the transformed disturbance $\boldsymbol{d}_\Delta(t)$. That is, $\gamma = 0$ means that the actual disturbance has to be constant $\boldsymbol{d}(t) = \bar{\boldsymbol{d}}$ and correctly estimated after event time t_k ($\bar{\boldsymbol{d}} = \hat{\boldsymbol{d}}_k$). In contrast, $\gamma = 2$ means that the disturbance estimate satisfies $\|\hat{\boldsymbol{d}}_k\| = d_{\max}$ for $t \geq t_k$, which likewise holds for the actual disturbance $\|\boldsymbol{d}(t)\| = \|\bar{\boldsymbol{d}}\| = d_{\max}$ but with opposite sign, i.e. $\hat{\boldsymbol{d}}_k = -\bar{\boldsymbol{d}}$ (Fig. 3.11). Note that without applying the disturbance estimation ($\hat{\boldsymbol{d}}_k = 0$ for all k), the range of γ reduces to $0 \leq \gamma \leq 1$.

An event is generated whenever the equation

$$\|\boldsymbol{x}_\Delta(t)\| = \left\| \int_{t_k}^{t} \mathrm{e}^{\boldsymbol{A}(t-\alpha)} \boldsymbol{E} \boldsymbol{d}_\Delta(\alpha) \, \mathrm{d}\alpha \right\| = \bar{e}$$

holds. The minimum inter-event time

$$T_{\min} = \min_k T_k$$

with $T_k = t_{k+1} - t_k$ ($k = 0, 1, 2, ...$) which satisfies this condition is given by

$$T_{\min} = \arg\min_t \left\{ \left\| \int_0^t \mathrm{e}^{\boldsymbol{A}(t-\alpha)} \boldsymbol{E} \boldsymbol{d}_\Delta(\alpha) \, \mathrm{d}\alpha \right\| = \bar{e} \right\}. \tag{3.39}$$

The following estimation yields a bound on this time. If the upper bound

$$\int_0^t \left\| \mathrm{e}^{\boldsymbol{A}\alpha} \boldsymbol{E} \right\| \, \mathrm{d}\alpha \cdot \gamma \, d_{\max} \geq \left\| \int_0^t \mathrm{e}^{\boldsymbol{A}(t-\alpha)} \boldsymbol{E} \boldsymbol{d}_\Delta(\alpha) \, \mathrm{d}\alpha \right\|$$

is set to $\bar{e}$, then the upper integral bound for which the relation

$$\bar{T} = \arg\min_t \left\{ \int_0^t \left\| \mathrm{e}^{\boldsymbol{A}\alpha} \boldsymbol{E} \right\| \, \mathrm{d}\alpha = \frac{\bar{e}}{\gamma \, d_{\max}} \right\} \tag{3.40}$$

holds is a lower bound of $T_{\min}$:

$$T_{\min} \geq \bar{T}. \tag{3.41}$$

Theorem 5. *For any bounded disturbance, the minimum inter-event time $T_{\min}$ of the event-based state-feedback loop* (3.1), (3.2), (3.12), (3.13), (3.21), (3.23), (3.24) *is bounded from below by $\bar{T}$ given by relation* (3.40).

This theorem highlights how the communication depends on the disturbances. This contrasts with discrete-time control, where the sampling frequency is chosen with respect to the plant properties (time constants) rather than the disturbance magnitude.

As the lower bound for the inter-event time decreases by decreasing the event threshold $\bar{e}$ and increases for large thresholds $\bar{e}$, Theorem 3 and Theorem 5 prove the statement given in Section 3.4.1 that a higher precision leads to a more frequent communication between the event generator and the control input generator and confirm the numerical results obtained in Example 2.

As $\gamma\, d_{\max}$ is a bound on $\boldsymbol{d}_\Delta(t)$ and not on the original disturbance $\boldsymbol{d}(t)$, the disturbance estimate $\hat{\boldsymbol{d}}_k$ reduces the transformed disturbance $\boldsymbol{d}_\Delta(t) = \boldsymbol{d}(t) - \hat{\boldsymbol{d}}_k$ for $\gamma < 1$, which ensures larger inter-event times T_k (left-hand side of Fig. 3.11).

If, on the other hand, $\gamma > 1$ holds, the estimate $\hat{\boldsymbol{d}}_k$ does not reduce the disturbance $\boldsymbol{d}_\Delta(t)$ in comparison to $\boldsymbol{d}(t)$. Instead, it may even increase the communication (right-hand side of the figure). In this case, the best estimate is $\hat{\boldsymbol{d}}_k = \boldsymbol{0}$, which means that no information about the disturbance should be used in the control input generator and the event generator because the information about the past disturbance does not say anything about the current disturbance.

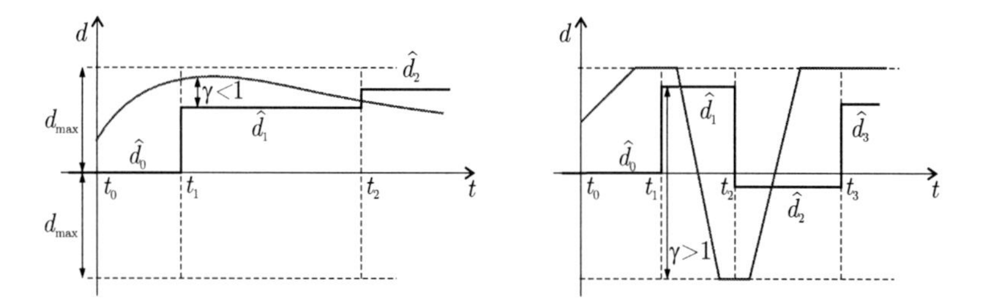

Figure 3.11.: Influence of the disturbance characteristic on the communication

Example 4 *Minimum inter-event time of the event-based controlled thermofluid process*

For the thermofluid process (2.7), (2.8), event condition (3.28) with $\bar{e} = 2$, $d_{\max} = 0.15$ *and* $\gamma = 1$, *the bound*

$$\bar{T} \approx 94\,\mathrm{s}$$

results. This bound is smaller than any inter-event time which occurs in Fig. 3.7 (Example 1), where

$$T_{\min} \approx 300\,\mathrm{s}$$

holds.

3.4.4. Small disturbances

This section shows that for any disturbance $\boldsymbol{d}(t)$ there exists a sufficiently small parameter $\bar{d}$ such that the event generator (3.21) does not generate any event besides the initialising event at time $t_0 = 0$.

The disturbance is represented in the following as $\boldsymbol{d}(t) = \bar{d}\tilde{\boldsymbol{d}}(t)$, where $\tilde{\boldsymbol{d}}(t)$ is an arbitrary finite vector function which satisfies

$$\|\tilde{\boldsymbol{d}}(t)\| \leq 1 \quad \text{for } t \geq 0$$

and $\bar{d}$ is a parameter determining the disturbance magnitude.

Lemma 5. *If the plant* (3.1), (3.2) *is stable, then for every bounded disturbance* $\boldsymbol{d}(t) = \bar{d}\tilde{\boldsymbol{d}}(t)$, *there exists a magnitude* $\bar{d}$ *such that the event generator does not generate any event for* $t > 0$. *The magnitude can be determined according to the relation*

$$|\bar{d}| < \frac{\bar{e}}{\int_0^\infty \|\mathrm{e}^{\boldsymbol{A}\alpha}\boldsymbol{E}\| \, \mathrm{d}\alpha} = \bar{d}_{\mathrm{UD}}. \tag{3.42}$$

Proof. See Appendix B.2, page 172. □

This result shows that in event-based control the communication is adapted to the severity of the disturbance. It can be seen that if the disturbance is small enough, no feedback is necessary to meet the performance requirements on the event-based control loop and, hence, the communication link between the sensor and the actuator is not used. Moreover, it gives a quantitative bound for the disturbance, for which no feedback occurs. In this situation the event-based control loop subject to the disturbance $\boldsymbol{d}(t)$ behaves similarly to the continuous-time state-feedback system without disturbances in the sense that the states $\boldsymbol{x}(t)$ and $\boldsymbol{x}_{\mathrm{s}}(t)$ of both systems do not deviate from one another more than the threshold $\bar{e}$:

$$\|\boldsymbol{x}(t) - \boldsymbol{x}_{\mathrm{s}}(t)\| < \bar{e}.$$

Considering the disturbance estimation, this result can be extended to large disturbances with small variation. If

$$\|\boldsymbol{d}_\Delta(t)\| = \|\boldsymbol{d}(t) - \hat{\boldsymbol{d}}_k\| < \bar{d}_{\mathrm{UD}}$$

holds for all times $t > t_k$, where $\hat{\boldsymbol{d}}_k$ is the disturbance estimate used by the control input generator after event time t_k, then no event is generated for all times $t > t_k$.

Note that if the plant is unstable no bound on the disturbance magnitude exists because then for any nonzero disturbance, which can be arbitrarily small, there exists a time t at which $\|\boldsymbol{x}_\Delta(t)\| = \bar{e}$ is satisfied (Eq. (3.20), [17]).

On the other hand, this result shows that if the disturbance satisfies Eq. (3.42), the disturbance is not detected by the event-based control as it causes no event. This affects the tracking property of the event-based state-feedback loop, which is investigated in the next section.

Example 5 *Undetected disturbances*

The preceding section has shown that no event is generated if the disturbance $\boldsymbol{d}(t)$ is small enough. Inequality (3.42) is now used to determine the maximum disturbance bound $\bar{d}_{\mathrm{UD}}$ for plant (2.7), (2.8) and event condition (3.28) with the event threshold $\bar{e} = 2$. It yields

$$\bar{d}_{\mathrm{UD}} \approx 0.0108.$$

The behaviour of the event-based control loop subject to a constant disturbance with amplitude $\bar{d} = 0.01$ is depicted in Fig. 3.10. It is shown that $|x_{\Delta,1}|$ becomes close to $\bar{e}$ but never reaches it. Thus, no event occurs due to the level behaviour which likewise holds for the temperature behaviour depicted in the right subplots of the figure.

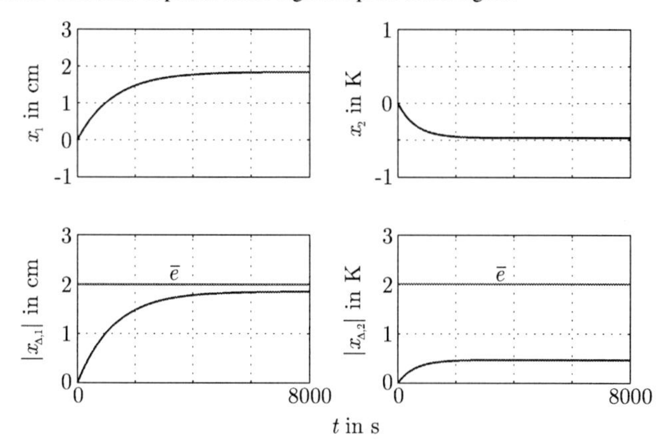

Figure 3.12.: *Undetected disturbance*

The bound obtained can also be substantiated by reconsidering the trajectories depicted in the middle subplots of Fig. 3.7 (Example 1). Here, after event time t_4 no further event is generated although the disturbance estimate $\hat{d}_4$ does not coincide with the disturbance magnitude $\bar{d}$ for $t \geq t_4$. However, the difference $\bar{d} - \hat{d}_4 \approx 0.003$ is small enough to satisfy inequality (3.42).

3.4.5. Setpoint tracking

To complement the previous results, this section investigates the setpoint tracking property of the event-based state-feedback loop for constant reference and disturbance signals:

$$\boldsymbol{w}(t) = \bar{\boldsymbol{w}}, \qquad \boldsymbol{d}(t) = \bar{\boldsymbol{d}}. \tag{3.43}$$

The control aim is to satisfy the requirement

$$\lim_{t \to \infty} \|\boldsymbol{y}(t) - \bar{\boldsymbol{w}}\| = 0 \tag{3.44}$$

for arbitrary setpoints $\bar{\boldsymbol{w}}$ and arbitrary disturbance magnitudes $\bar{\boldsymbol{d}}$. In a relaxed form, this aim can be formulated as

$$\lim_{t \to \infty} d(\boldsymbol{y}(t), \Omega_{\mathrm{y}}(\bar{\boldsymbol{w}})) = 0 \tag{3.45}$$

where $\Omega_{\mathrm{y}}(\bar{\boldsymbol{w}})$ is the set of acceptable output values around the desired setpoint $\bar{\boldsymbol{w}}$ and the term $d(\boldsymbol{y}(t), \Omega_{\mathrm{y}}(\bar{\boldsymbol{w}}))$ denotes the distance between the output $\boldsymbol{y}(t)$ and the set $\Omega_{\mathrm{y}}(\bar{\boldsymbol{w}})$ according to

$$d(\boldsymbol{y}(t), \Omega_{\mathrm{y}}(\bar{\boldsymbol{w}})) := \min_{\hat{\boldsymbol{y}} \in \Omega_{\mathrm{y}}(\bar{\boldsymbol{w}})} \|\boldsymbol{y}(t) - \hat{\boldsymbol{y}}\|.$$

Lemma 6 and Theorem 6 summarise the tracking properties of the undisturbed event-based control loop and the event-based state feedback affected by a constant disturbance. The corresponding proofs can be found in the appendix (see Appendices B.3, B.4, page 172).

Lemma 6. *For $\boldsymbol{d}(t) = \mathbf{0}$ and $\boldsymbol{w}(t) = \bar{\boldsymbol{w}}$, the event-based state-feedback loop* (3.1), (3.2), (3.12), (3.13), (3.21), (3.23), (3.24) *possesses the setpoint tracking property* (3.44).

Theorem 6. *For constant exogenous signals* (3.43) *and a stable plant* (3.1), (3.2), *the event-based state-feedback loop possesses the setpoint tracking property* (3.45) *with*

$$\Omega_{\mathrm{y}}(\bar{\boldsymbol{w}}) = \{\boldsymbol{y} : \|\boldsymbol{y} - \bar{\boldsymbol{w}}\| \leq e_{\mathrm{ys}}\},$$

where the bound e_{ys} depends upon the disturbance $\bar{\boldsymbol{d}}$ as follows:

$$e_{\mathrm{ys}} = \max\{\|\boldsymbol{C}\boldsymbol{A}^{-1}\boldsymbol{E}\bar{\boldsymbol{d}}\|, \|\boldsymbol{C}\bar{\boldsymbol{A}}^{-1}\boldsymbol{E}\bar{\boldsymbol{d}}\|\}.$$

If the plant is unstable, this bound reduces to

$$e_{\mathrm{yus}} = \|\boldsymbol{C}\bar{\boldsymbol{A}}^{-1}\boldsymbol{E}\bar{\boldsymbol{d}}\|.$$

Theorem 6 shows that setpoint tracking occurs for the disturbed system only under specific conditions. If the disturbance is small enough not to bring about an event, setpoint tracking (3.44) results if

$$\boldsymbol{C}\boldsymbol{A}^{-1}\boldsymbol{E}\bar{\boldsymbol{d}} = 0$$

holds. If the disturbance is large enough to cause an event, tracking occurs if

$$\boldsymbol{C}\bar{\boldsymbol{A}}^{-1}\boldsymbol{E}\bar{\boldsymbol{d}} = 0$$

holds. Both conditions on the disturbance magnitude $\bar{\boldsymbol{d}}$ are, however, very restrictive (see Example 1).

Note that the bound e_{ys} does not depend upon the event threshold $\bar{e}$. The reason for this is given by the fact that the deviation $\boldsymbol{y}(t) - \bar{\boldsymbol{w}}$ is not brought about by the event-based character of the loop but by the fact that the continuous-time state-feedback loop (3.5), (3.6) likewise has this control error. To overcome this problem a dynamical controller is incorporated into the control input generator in Section 4.3.

3.4.6. Robustness

In the previous analysis it was always assumed that the plant model is accurate in the sense that parameter uncertainties can be neglected. However, in practice, model uncertainties generally occur. In this case, the models (3.12), (3.13) used in the control input generator and in the event generator may differ from the actual plant dynamics [11, 20]. The consequences on the performance of the event-based state-feedback loop are investigated in this section.

The plant is assumed to have unknown parameter uncertainties

$$\begin{aligned} \dot{\boldsymbol{x}}(t) &= (\boldsymbol{A} + \Delta\boldsymbol{A})\boldsymbol{x}(t) + (\boldsymbol{B} + \Delta\boldsymbol{B})\boldsymbol{u}(t) + (\boldsymbol{E} + \Delta\boldsymbol{E})\boldsymbol{d}(t), \quad \boldsymbol{x}(0) = \boldsymbol{x}_0 \qquad (3.46) \\ \boldsymbol{y}(t) &= (\boldsymbol{C} + \Delta\boldsymbol{C})\boldsymbol{x}(t), \qquad (3.47) \end{aligned}$$

which are bounded according to

$$\begin{aligned} \|\Delta\boldsymbol{A}\| &\leq \Delta A_{\max} \qquad (3.48) \\ \|\Delta\boldsymbol{B}\| &\leq \Delta B_{\max} \\ \|\Delta\boldsymbol{E}\| &\leq \Delta E_{\max} \\ \|\Delta\boldsymbol{C}\| &\leq \Delta C_{\max}. \end{aligned}$$

With the state feedback

$$\boldsymbol{u}(t) = -\boldsymbol{K}\boldsymbol{x}(t),$$

the continuous-time state-feedback loop

$$\dot{\boldsymbol{x}}_{\mathrm{CT}}(t) = \underbrace{(\boldsymbol{A} + \Delta\boldsymbol{A} - (\boldsymbol{B} + \Delta\boldsymbol{B})\boldsymbol{K})}_{\bar{\boldsymbol{A}}_{\mathrm{U}}} \boldsymbol{x}_{\mathrm{CT}}(t) + (\boldsymbol{E} + \Delta\boldsymbol{E})\boldsymbol{d}(t), \quad \boldsymbol{x}_{\mathrm{CT}}(0) = \boldsymbol{x}_0 \quad (3.49)$$

$$\boldsymbol{y}_{\mathrm{CT}}(t) = (\boldsymbol{C} + \Delta\boldsymbol{C})\boldsymbol{x}_{\mathrm{CT}}(t) \quad (3.50)$$

results. The controller matrix $\boldsymbol{K}$ is assumed to be designed so that $\bar{\boldsymbol{A}}_{\mathrm{U}}$ is Hurwitz for any uncertainties satisfying the bounds (3.48) (for the design of continuous-time robust controllers the reader is referred to [24, 79]).

Behaviour of the event-based control loop. With the plant (3.46), (3.47) and the control input generator (3.12), (3.13), the state behaviour of the event-based state-feedback loop can be described for the time interval $[t_k, t_{k+1})$ by the state-space model

$$\begin{pmatrix} \dot{\boldsymbol{x}}(t) \\ \dot{\boldsymbol{x}}_{\mathrm{s}}(t) \end{pmatrix} = \begin{pmatrix} \boldsymbol{A} + \Delta\boldsymbol{A} & -(\boldsymbol{B} + \Delta\boldsymbol{B})\boldsymbol{K} \\ \boldsymbol{O} & \bar{\boldsymbol{A}} \end{pmatrix} \begin{pmatrix} \boldsymbol{x}(t) \\ \boldsymbol{x}_{\mathrm{s}}(t) \end{pmatrix} + \begin{pmatrix} \boldsymbol{E} + \Delta\boldsymbol{E} \\ \boldsymbol{O} \end{pmatrix} \boldsymbol{d}(t) + \begin{pmatrix} \boldsymbol{O} \\ \boldsymbol{E} \end{pmatrix} \hat{\boldsymbol{d}}_k$$

with the initial condition

$$\begin{pmatrix} \boldsymbol{x}(t_k) \\ \boldsymbol{x}_{\mathrm{s}}(t_k^+) \end{pmatrix} = \begin{pmatrix} \boldsymbol{x}(t_k) \\ \boldsymbol{x}(t_k) \end{pmatrix}.$$

Using state transformation (3.14), the transformed model

$$\begin{pmatrix} \dot{\boldsymbol{x}}_{\Delta}(t) \\ \dot{\boldsymbol{x}}_{\mathrm{s}}(t) \end{pmatrix} = \begin{pmatrix} \boldsymbol{A} + \Delta\boldsymbol{A} & \Delta\boldsymbol{A} - \Delta\boldsymbol{B}\boldsymbol{K} \\ \boldsymbol{O} & \bar{\boldsymbol{A}} \end{pmatrix} \begin{pmatrix} \boldsymbol{x}_{\Delta}(t) \\ \boldsymbol{x}_{\mathrm{s}}(t) \end{pmatrix} + \begin{pmatrix} \boldsymbol{E} + \Delta\boldsymbol{E} \\ \boldsymbol{O} \end{pmatrix} \boldsymbol{d}(t) + \begin{pmatrix} -\boldsymbol{E} \\ \boldsymbol{E} \end{pmatrix} \hat{\boldsymbol{d}}_k$$

$$\begin{pmatrix} \boldsymbol{x}_{\Delta}(t_k^+) \\ \boldsymbol{x}_{\mathrm{s}}(t_k^+) \end{pmatrix} = \begin{pmatrix} 0 \\ \boldsymbol{x}(t_k) \end{pmatrix}$$

results. In comparison to the state-space model (3.15) obtained for negligible model uncertainties, the upper right element $(\Delta\boldsymbol{A} - \Delta\boldsymbol{B}\boldsymbol{K})$ is nonzero. Thus, a strict decomposition of the state $\boldsymbol{x}(t)$ into the difference state $\boldsymbol{x}_\Delta(t)$ and the model state $\boldsymbol{x}_\mathrm{s}(t)$ (cf. Section 3.3.2, Fig. 3.3) cannot be achieved because $\boldsymbol{x}_\Delta(t)$ depends additionally on the model state $\boldsymbol{x}_\mathrm{s}(t)$, which is illustrated in Fig. 3.13.

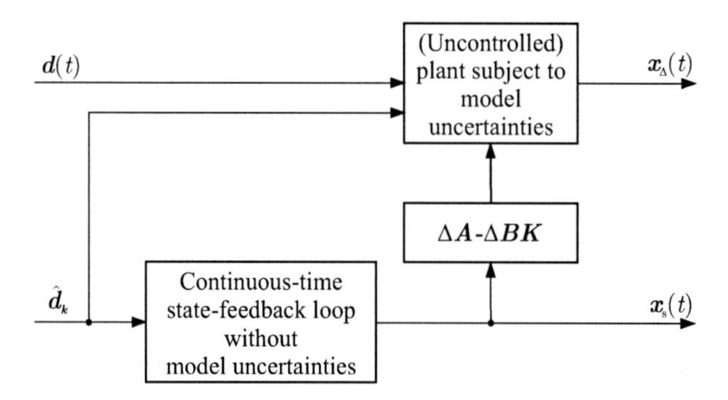

Figure 3.13.: Decomposition of the state trajectory $\boldsymbol{x}(t) = \boldsymbol{x}_\Delta(t) + \boldsymbol{x}_\mathrm{s}(t)$

The state trajectories can be described by

$$\boldsymbol{x}_\mathrm{s}(t) = \mathrm{e}^{\bar{\boldsymbol{A}}(t-t_k)}\boldsymbol{x}(t_k) + \bar{\boldsymbol{A}}^{-1}\left(\mathrm{e}^{\bar{\boldsymbol{A}}(t-t_k)} - \boldsymbol{I}_n\right)\boldsymbol{E}\hat{\boldsymbol{d}}_k \tag{3.51}$$

$$\begin{aligned}\boldsymbol{x}_\Delta(t) &= \int_{t_k}^{t} \mathrm{e}^{(\boldsymbol{A}+\Delta\boldsymbol{A})(t-\alpha)}((\boldsymbol{E}+\Delta\boldsymbol{E})\boldsymbol{d}(\alpha) - \boldsymbol{E}\hat{\boldsymbol{d}}_k)\,\mathrm{d}\alpha \\ &\quad + \int_{t_k}^{t} \mathrm{e}^{(\boldsymbol{A}+\Delta\boldsymbol{A})(t-\alpha)}(\Delta\boldsymbol{A} - \Delta\boldsymbol{B}\boldsymbol{K})\boldsymbol{x}_\mathrm{s}(\alpha)\,\mathrm{d}\alpha.\end{aligned} \tag{3.52}$$

Again, the first equation corresponds to the state trajectory of the continuous-time state-feedback loop (3.5), (3.6) without model uncertainties, whereas the second equation shows the consequences of model uncertainties on the event-based state feedback. The consequences are the following:

- The disturbance estimation (3.24) generally provides imprecise disturbance estimates in the case of model uncertainties because the difference between the model state $\boldsymbol{x}_\mathrm{s}(t)$ and the plant state $\boldsymbol{x}(t)$ is not only brought about by the disturbance but also by the difference of the model and the plant dynamics. However, an analytic expression to correctly estimate the disturbance based on Eq. (3.52) can not be obtained as the exact model uncertainties are usually unknown.

- For the determination of the minimum inter-event time (Section 3.4.3), additional information about the model state $\boldsymbol{x}_{\mathrm{s}}(t)$ are required. This likewise holds for the disturbance bound $\bar{d}_{\mathrm{UD}}$ (Section 3.4.4).

- For small model uncertainties ($\Delta\boldsymbol{A} = \Delta\boldsymbol{B} = \Delta\boldsymbol{E} \approx \boldsymbol{O}$), Eq. (3.52) reduces to Eq. (3.20).

Stability analysis. For the stability analysis, the state behaviour of the event-based control loop (3.13), (3.21), (3.46) and the state behaviour of the continuous-time state-feedback loop (3.49) are compared.

The behaviour of the approximation error $\boldsymbol{e}(t) = \boldsymbol{x}(t) - \boldsymbol{x}_{\mathrm{CT}}(t)$ is described by the state-space model

$$\begin{aligned}
\dot{\boldsymbol{e}}(t) &= \dot{\boldsymbol{x}}(t) - \dot{\boldsymbol{x}}_{\mathrm{CT}}(t) \\
&= (\boldsymbol{A} + \Delta\boldsymbol{A})\boldsymbol{x}(t) - (\boldsymbol{B} + \Delta\boldsymbol{B})\boldsymbol{K}\boldsymbol{x}_{\mathrm{s}}(t) + (\boldsymbol{E} + \Delta\boldsymbol{E})\boldsymbol{d}(t) \\
&\quad -(\boldsymbol{A} + \Delta\boldsymbol{A} - (\boldsymbol{B} + \Delta\boldsymbol{B})\boldsymbol{K})\boldsymbol{x}_{\mathrm{CT}}(t) + (\boldsymbol{E} + \Delta\boldsymbol{E})\boldsymbol{d}(t) \\
&= \bar{\boldsymbol{A}}_{\mathrm{U}}\boldsymbol{e}(t) + (\boldsymbol{B} + \Delta\boldsymbol{B})\boldsymbol{K}\boldsymbol{x}_{\Delta}(t), \quad \boldsymbol{e}(0) = \boldsymbol{0}.
\end{aligned}$$

As $\bar{\boldsymbol{A}}_{\mathrm{U}}$ is assumed to be Hurwitz, $\Delta\boldsymbol{B}$ is bounded according to Eq. (3.48) and $\|\boldsymbol{x}_{\Delta}(t)\|$ is bounded by $\bar{e}$ according to Eq. (3.21), the maximum error bound is given by

$$\|\boldsymbol{e}(t)\| \leq e_{\mathrm{max,U}} = \bar{e} \cdot \int_0^{\infty} \max_{\Delta\boldsymbol{A},\, \Delta\boldsymbol{B}} \left\| \mathrm{e}^{\bar{\boldsymbol{A}}_{\mathrm{U}}\alpha}(\boldsymbol{B} + \Delta\boldsymbol{B})\boldsymbol{K} \right\| \, \mathrm{d}\alpha. \tag{3.53}$$

Theorem 7. *If the state $\boldsymbol{x}_{\mathrm{CT}}(t)$ of the continuous-time state-feedback loop* (3.49), (3.50) *is GUUB for any bounded model uncertainties $\|\Delta\boldsymbol{A}\| \leq \Delta A_{\mathrm{max}}$, $\|\Delta\boldsymbol{B}\| \leq \Delta B_{\mathrm{max}}$, then the state $\boldsymbol{x}(t)$ of event-based state-feedback loop* (3.12), (3.13), (3.21), (3.46), (3.47) *is GUUB and the approximation error $\boldsymbol{e}(t) = \boldsymbol{x}(t) - \boldsymbol{x}_{\mathrm{CT}}(t)$ is bounded from above by*

$$\|\boldsymbol{e}(t)\| \leq e_{\mathrm{max,U}}$$

with $e_{\mathrm{max,U}}$ given by Eq. (3.53).

This result shows that if the controller $\boldsymbol{K}$ is chosen so that the continuous-time state-feedback loop is stable for any model uncertainties (3.48), the event-based state-feedback loop is stable as well. Note that the approximation accuracy can be arbitrarily adjusted through the event threshold $\bar{e}$. However, the computational complexity for numerically deriving $e_{\max,\mathrm{U}}$ is significantly increased.

Remark. The minimum time interval $T_{\min}$ between two consecutive events is given by

$$\begin{aligned} T_{\min} &= \arg\min_t \left\{ \left\| \int_0^t \mathrm{e}^{(\boldsymbol{A}+\Delta\boldsymbol{A})(t-\alpha)}((\boldsymbol{E}+\Delta\boldsymbol{E})\,\boldsymbol{d}(\alpha) - \boldsymbol{E}\hat{\boldsymbol{d}}_k)\,\mathrm{d}\alpha \right.\right. \\ &\qquad \left.\left. + \int_0^t \mathrm{e}^{(\boldsymbol{A}+\Delta\boldsymbol{A})(t-\alpha)}(\Delta\boldsymbol{A} - \Delta\boldsymbol{B}\boldsymbol{K})\boldsymbol{x}_{\mathrm{s}}(\alpha)\,\mathrm{d}\alpha \right\| = \bar{e} \right\} \end{aligned}$$

according to Eqs. (3.21), (3.52). To derive a lower bound on the minimum inter-event time ($\bar{T} \leq T_{\min}$) according to Section 3.4.3, an upper bound $\max_{t\in[0,\infty)} \|\boldsymbol{x}_{\mathrm{s}}(t)\|$ of the model state $\boldsymbol{x}_{\mathrm{s}}(t)$ is required. Applying Theorem 7, an upper bound can be determined as follows:

$$\begin{aligned} \max_{t\in[0,\infty)} \|\boldsymbol{x}_{\mathrm{s}}(t)\| &= \max_{t\in[0,\infty)} \|\boldsymbol{x}(t) - \boldsymbol{x}_{\Delta}(t)\| \\ &= \max_{t\in[0,\infty)} \|\boldsymbol{x}_{\mathrm{CT}}(t) + \boldsymbol{e}(t)\| + \bar{e} \\ &= x_{\max,\mathrm{CT}} + e_{\max,\mathrm{U}} + \bar{e} \end{aligned}$$

with

$$x_{\max,\mathrm{CT}} = \max_{t\in[0,\infty)} \|\boldsymbol{x}_{\mathrm{CT}}(t)\|.$$

Apparently, this procedure is very conservative and may generally lead to small bounds on the inter-event times. However, it can be expected that model uncertainties significantly affect the communication of the event-based state-feedback loop. To better deal with model uncertainties in terms of reducing the communication, Section 4.4 presents a modification of the disturbance estimation (3.23), (3.24).

The influence of model uncertainties on the stability and communication properties of the event-based control loop is evaluated by experiments in Chapter 6.

3.5. Discrete-time state feedback

3.5.1. Model

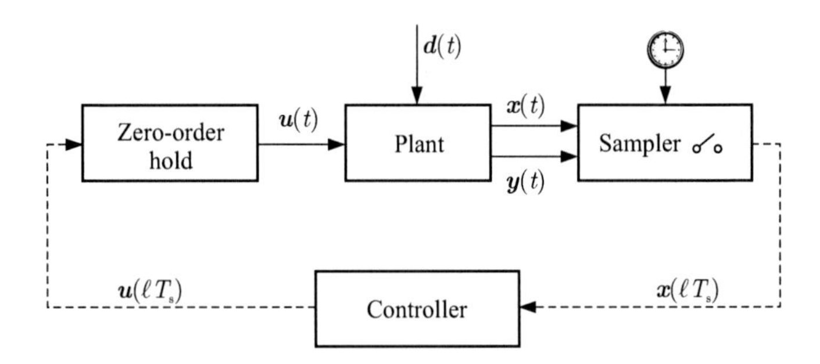

Figure 3.14.: Discrete-time control loop

Discrete-time control [32] represents the traditional control implementation on digital hardware. Here, instead of the event generator a time-driven sampler is used in the control loop to invoke an equidistant sampling (Fig. 3.14). This section aims at comparing the discrete-time state-feedback loop with the corresponding continuous-time state-feedback loop equivalently to the previous analysis. The results are later used to compare the performance guarantees of event-based state feedback and discrete-time state feedback.

Considered is the plant (3.1), (3.2) and the controller

$$\boldsymbol{u}(\ell T_{\mathrm{s}}) = -\boldsymbol{K}_{\mathrm{DT}}\boldsymbol{x}(\ell T_{\mathrm{s}}), \quad t \in [\ell T_{\mathrm{s}}, (\ell+1)T_{\mathrm{s}}), \tag{3.54}$$

where "DT" is used from now on to indicate signals and specific parameters of the discrete-time state-feedback loop.

Similar to the description of the control input generation in Section 3.3.1 for the event-based control loop, the discrete-time controller and the zero-order hold are first lumped together, which yields the state-space model

$$\begin{aligned}
\dot{\boldsymbol{x}}_{\mathrm{s,DT}}(t) &= \mathbf{0} \\
\boldsymbol{x}_{\mathrm{s,DT}}(\ell T_{\mathrm{s}}^{+}) &= \boldsymbol{x}(\ell T_{\mathrm{s}}) \\
\boldsymbol{u}(\ell T_{\mathrm{s}}) &= -\boldsymbol{K}_{\mathrm{DT}}\boldsymbol{x}_{\mathrm{s,DT}}(t)
\end{aligned}$$

with ℓT_{s}^{+} denoting the time instance immediately after the update of the *controller state* $\boldsymbol{x}_{\mathrm{s,DT}}$

with the measured state $\boldsymbol{x}(\ell T_s)$ at the sampling time ℓT_s.

By introducing the difference state

$$\boldsymbol{x}_{\Delta,\mathrm{DT}}(t) = \boldsymbol{x}_{\mathrm{DT}}(t) - \boldsymbol{x}_{\mathrm{s,DT}}(t) \tag{3.55}$$

where $\boldsymbol{x}_{\mathrm{DT}}(t)$ is used to denote the state of plant (3.1), (3.2) in the discrete-time state-feedback loop also referred to as the state of the discrete-time state-feedback loop in the following, the discrete-time control loop can be described by

$$\dot{\boldsymbol{x}}_{\mathrm{DT}}(t) = (\boldsymbol{A} - \boldsymbol{B}\boldsymbol{K}_{\mathrm{DT}})\boldsymbol{x}_{\mathrm{DT}}(t) + \boldsymbol{B}\boldsymbol{K}_{\mathrm{DT}}\boldsymbol{x}_{\Delta,\mathrm{DT}}(t) + \boldsymbol{E}\boldsymbol{d}(t), \ \ \boldsymbol{x}(0) = \boldsymbol{x}_0 \tag{3.56}$$

$$\boldsymbol{y}_{\mathrm{DT}}(t) = \boldsymbol{C}\boldsymbol{x}_{\mathrm{DT}}(t). \tag{3.57}$$

Note that although there are specific methods to suitably design the controller in the discrete-time case [83], it is assumed in the subsequent analysis that the controller matrix $\boldsymbol{K}$ of the continuous-time control loop and the controller matrix $\boldsymbol{K}_{\mathrm{DT}}$ of the discrete-time control loop are identical ($\boldsymbol{K} = \boldsymbol{K}_{\mathrm{DT}}$) and chosen so that the states of both control loops are ultimately bounded.

A condition for proving the ultimate boundedness of the discrete-time control loop (3.56), (3.57) is stated in Theorem 19, page 175.

3.5.2. Comparison of the discrete-time state-feedback loop and the continuous-time state-feedback loop

The approximation error $\boldsymbol{e}_{\mathrm{DT}}(t) = \boldsymbol{x}_{\mathrm{DT}}(t) - \boldsymbol{x}_{\mathrm{CT}}(t)$ between the state $\boldsymbol{x}_{\mathrm{DT}}(t)$ of the discrete-time state-feedback loop (3.56), (3.57) and the state $\boldsymbol{x}_{\mathrm{CT}}(t)$ of the continuous-time state-feedback loop (3.5), (3.6) ($\boldsymbol{w}(t) = \boldsymbol{0}$) is given by

$$\begin{aligned}
\dot{\boldsymbol{e}}_{\mathrm{DT}}(t) &= \dot{\boldsymbol{x}}_{\mathrm{DT}}(t) - \dot{\boldsymbol{x}}_{\mathrm{CT}}(t) \\
&= (\boldsymbol{A} - \boldsymbol{B}\boldsymbol{K})\boldsymbol{x}_{\mathrm{DT}}(t) + \boldsymbol{B}\boldsymbol{K}\boldsymbol{x}_{\Delta,\mathrm{DT}}(t) + \boldsymbol{E}\boldsymbol{d}(t) - (\boldsymbol{A} - \boldsymbol{B}\boldsymbol{K})\boldsymbol{x}_{\mathrm{CT}}(t) - \boldsymbol{E}\boldsymbol{d}(t) \\
&= \bar{\boldsymbol{A}}\boldsymbol{e}_{\mathrm{DT}}(t) + \boldsymbol{B}\boldsymbol{K}\boldsymbol{x}_{\Delta,\mathrm{DT}}(t), \quad \boldsymbol{e}_{\mathrm{DT}}(0) = \boldsymbol{0}.
\end{aligned}$$

This differential equation is used to show that the state $\boldsymbol{x}_{\mathrm{DT}}(t)$ of the discrete-time state-feedback loop remains in the bounded surrounding

$$\boldsymbol{x}_{\mathrm{DT}}(t) \in \Omega_{\mathrm{e,DT}}(\boldsymbol{x}_{\mathrm{CT}}(t)) = \{\boldsymbol{x}_{\mathrm{DT}} : \|\boldsymbol{x}_{\mathrm{DT}} - \boldsymbol{x}_{\mathrm{CT}}(t)\| \leq e_{\mathrm{max,DT}}\}, \quad \forall t$$

of the state $\boldsymbol{x}_{\mathrm{CT}}(t)$ of the continuous-time state-feedback loop, which is summarised in the following result.

Theorem 8. *Under the assumption $\|\bar{\boldsymbol{A}}_{\mathrm{DT}}\| < 1$, the approximation error*

$$\boldsymbol{e}_{\mathrm{DT}}(t) = \boldsymbol{x}_{\mathrm{DT}}(t) - \boldsymbol{x}_{\mathrm{CT}}(t)$$

between the state $\boldsymbol{x}_{\mathrm{DT}}(t)$ of the discrete-time state-feedback loop (3.56), (3.57) *and the state $\boldsymbol{x}_{\mathrm{CT}}(t)$ of the continuous-time state-feedback loop* (3.5), (3.6) *is bounded from above by*

$$\|\boldsymbol{e}_{\mathrm{DT}}(t)\| \leq e_{\max,\mathrm{DT}} = x_{\Delta\max} \cdot \int_0^{\infty} \left\| \mathrm{e}^{\bar{\boldsymbol{A}}\alpha} \boldsymbol{B}\boldsymbol{K} \right\| \, \mathrm{d}\alpha \tag{3.58}$$

with

$$x_{\Delta\max} = \int_0^{T_s} \left\| \mathrm{e}^{\boldsymbol{A}\alpha} \right\| \, \mathrm{d}\alpha \cdot \left(\|\bar{\boldsymbol{A}}\| x_{\mathrm{DT},\max} + \|\boldsymbol{E}\| d_{\max} \right) \tag{3.59}$$

and

$$\begin{aligned}
x_{\mathrm{DT},\max} &= \max_{\ell \in \{0,1,\ldots,\infty\}} \left\| \bar{\boldsymbol{A}}_{\mathrm{DT}}^{\ell} \right\| \cdot \|\boldsymbol{x}_0\| + \frac{\|\boldsymbol{E}_{\mathrm{DT}}\| \, d_{\max}}{1 - \|\bar{\boldsymbol{A}}_{\mathrm{DT}}\|} \\
\bar{\boldsymbol{A}}_{\mathrm{DT}} &= \boldsymbol{A}_{\mathrm{DT}} - \boldsymbol{B}_{\mathrm{DT}} \boldsymbol{K} \\
\boldsymbol{A}_{\mathrm{DT}} &= \mathrm{e}^{\boldsymbol{A} T_{\mathrm{s}}} \\
\boldsymbol{B}_{\mathrm{DT}} &= \int_0^{T_s} \mathrm{e}^{\boldsymbol{A}\alpha} \boldsymbol{B} \, \mathrm{d}\alpha \\
\boldsymbol{E}_{\mathrm{DT}} &= \int_0^{T_s} \mathrm{e}^{\boldsymbol{A}\alpha} \boldsymbol{E} \, \mathrm{d}\alpha.
\end{aligned}$$

Proof. See Appendix B.5, page 174. □

The theorem brings out that the maximum approximation error of the discrete-time state feedback depends on the disturbance $\boldsymbol{d}(t)$ and the sampling period T_{s}, where $x_{\Delta\max}$ can be chosen arbitrarily by changing the sampling period T_{s}. Hence, analogically to the event-based controller (cf. Section 3.4.1), the discrete-time controller can be made to mimic a continuous-time state feedback with arbitrary precision by accordingly choosing the sampling period T_{s}.

Example 6 *Comparison of event-based control and discrete-time control*

In this example, the stable scalar system

$$\dot{x}(t) = -0.5x(t) + u(t) + d(t), \quad x(0) = 0$$

with the controller

$$u(t) = -1.5x(t)$$

is investigated. The disturbance $d(t)$ affecting the system is assumed to be bounded with $|d(t)| \leq d_{\max} = 1$. The sampling period T_s of the discrete-time state-feedback loop is set to $T_s = 0.25$ s, whereas the event threshold $\bar{e}$ has been varied in order to get comparable results. The disturbance parameter γ is assumed to be $\gamma = 1$ (no disturbance estimation).

Table 3.1.: Guaranteed bounds

Control scheme	T_s	$\bar{e}$	$e_{\max}$	$\bar{T}$
Discrete-time state feedback	0.25 s	–	0.35	0.25 s
Event-based state feedback	–	0.47	0.35	0.54 s
Event-based state feedback	–	0.24	0.18	0.25 s

Table 3.1 summarises the results obtained by applying Eqs. (3.32), (3.40) and (3.58). It can be seen that, for the event threshold $\bar{e} = 0.47$, the event-based state-feedback loop guarantees the same performance bound as the discrete-time control (identical $e_{\max}$) but the minimum inter-event time is two times larger than the sampling period T_s. Vice versa, for $\bar{e} = 0.24$, the event-based state feedback has at most the communication of the discrete-time state feedback ($T_s = \bar{T}$) but with a significantly smaller error bound $e_{\max}$. A thorough comparison of these schemes can be found in [13].

Note that the analysis only gives performance guarantees. Therefore, the actual performance of the discrete-time control loop might be better than the actual performance of the event-based control loop even for the bounds specified in Tab. 3.1. In the discrete-time case, the analysis may be improved by e.g. reducing the conservatism introduced through Eq. (3.59).

4. Extensions

This chapter presents extensions of the event-based state-feedback scheme which are mainly focussed on improving its practical applicability. The extensions include the incorporation of a state observer (Section 4.2), a dynamical controller (Section 4.3), an improved disturbance estimation (Section 4.4), and the discrete-time implementation of the control input generator and the event generator (Section 4.5).

4.1. Limitations of the event-based state feedback

The event-based state feedback introduced in the previous chapter relies on the availability of the full state information and uses a proportional controller to generate the control input $\boldsymbol{u}(t)$. From a practical perspective these prerequisites are often inadequate as the measurable output $\boldsymbol{y}(t)$ may distinguish from the plant state $\boldsymbol{x}(t)$ and the proportional controller is only capable to guarantee setpoint tracking under very restrictive conditions (see Section 3.4.5).

To overcome these problems, the components of the event-based control loop (Fig. 3.1) have to be modified as discussed in this chapter.

- For the case of a non-measurable state $\boldsymbol{x}(t)$, Section 4.2 presents an event-based output-feedback control, where a state observer is included in the event generator in order to determine an approximate state $\hat{\boldsymbol{x}}(t)$ of the plant state $\boldsymbol{x}(t)$ based on the measured output $\boldsymbol{y}(t)$ of the plant.

- In Section 4.3, the control input generator is extended by incorporating a dynamical controller. The analysis concentrates on the setpoint tracking properties of the extended event-based control loop for constant and time-varying disturbances.

- To improve the disturbance estimation in terms of compensating model uncertainties, Section 4.4 introduces an augmented disturbance estimate. Moreover, to relax the requirement of the inverse system matrix $\boldsymbol{A}^{-1}$ (Section 3.3.4), this section also proposes an alternative disturbance estimator which uses a disturbance observer.

- Both the control input generator and the event generator are generally implemented on digital hardware and usually operate with a fixed sampling time T_{s}. As the main consequence, a continuous-time monitoring of the event condition according to Eq. (3.21) is not feasible. The effect of this circumstance on the behaviour of the event-based state-feedback loop is investigated in Section 4.5.

Except for the extensions of the disturbance estimator discussed in Section 4.4, which do not affect the stability of the event-based state-feedback loop, the analysis in this chapter is focussed on investigating the stability properties of the event-based control loop and the determination of the approximation error bounds. Moreover, in all these situations, the influence of the extensions on the information exchange over the feedback link in the event-based control loop is investigated.

4.2. Event-based output feedback

This section presents a method for *event-based output feedback*, where the full state information $\boldsymbol{x}(t)$ is assumed to be non-measurable and the measurable output $\boldsymbol{y}(t)$ is affected by measurement noise $\boldsymbol{v}(t)$. The event-based output-feedback loop is depicted in Fig. 4.1.

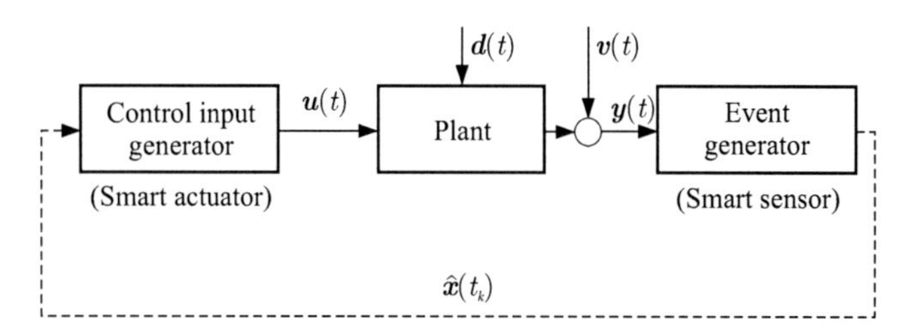

Figure 4.1.: Event-based output-feedback loop

The approach investigated in this section and proposed in [7] extends the event-based state feedback by incorporating a Luenberger state observer in the event generator which continuously gets the measurable output $\boldsymbol{y}(t)$. It distinguishes from the related literature as follows:

- References [46, 72] do not incorporate a state observer. Instead, the output information $\boldsymbol{y}(t_k)$ is directly processed in order to update the input signal $\boldsymbol{u}(t_k)$ at event times t_k $(k = 0, 1, 2, ...)$.
- The approaches in [28, 77, 107, 119] use a modified Kalman filter to reconstruct the plant state $\boldsymbol{x}(t)$. The observer operates in an event-based way as it is provided with the measurable output $\boldsymbol{y}(t)$ only at event time t_k.

In summary, the event-based state feedback is modified in this section by

- incorporating a state observer in the event generator,
- defining an adapted event condition, and
- sending the observer state $\hat{\boldsymbol{x}}(t_k)$ at event times t_k.

As the observer state $\hat{\boldsymbol{x}}(t)$ generally differs from the actual plant state $\boldsymbol{x}(t)$, i.e. $\hat{\boldsymbol{x}}(t) \neq \boldsymbol{x}(t)$, the event-based output feedback necessitates a considerable change of the previous analysis. The main results of the following analysis are:

1. The observation error $\hat{x}_\Delta(t) = x(t) - \hat{x}(t)$ is bounded when incorporating a state observer in the event generator according to Fig. 4.2 (Lemma 7).

2. By defining a new event triggering mechanism, which uses the observer state $\hat{x}(t)$, a stable behaviour of the event-based output-feedback loop subject to disturbances and measurement noise can be guaranteed and a bound on the approximation error can be derived (Theorem 9).

3. There exists a minimum inter-event time (Theorem 10).

Problem formulation. In the event-based state feedback, the event generator as well as the control input generator use the model (3.12), (3.13) between two consecutive events ($t \in [t_k, t_{k+1})$) both for the input generation and for the event generation. Here, events are generated based on Eq. (3.21) if

$$\|x(t) - x_s(t)\| = \bar{e}$$

holds. At event time t_k, the state information $x(t_k)$ is sent from the event generator towards the control input generator and the model state $x_s(t)$ is updated with the measured state $x(t_k)$ both in the control input generator and the event generator.

Apparently, if the state $x(t)$ cannot be measured, an update of the model states $x_s(t)$ according to Eq. (3.12) and an event generation according to Eq. (3.21) is not possible. Hence, the control input generator and the event generator have to be modified as depicted in Fig. 4.2 and described in the following.

4.2.1. Description of the components

Plant. Throughout this section, the plant is given by

$$\dot{x}(t) = Ax(t) + Bu(t) + Ed(t), \quad x(0) = x_0 \tag{4.1}$$

$$y(t) = Cx(t) + v(t), \tag{4.2}$$

where the disturbance $d \in \mathbb{R}^l$ and the measurement noise $v \in \mathbb{R}^r$ are assumed to be bounded:

$$\|d(t)\| \leq d_{\max}, \quad \|v(t)\| \leq v_{\max}.$$

The pair (A, B) is assumed to be controllable and the pair (A, C) to be observable.

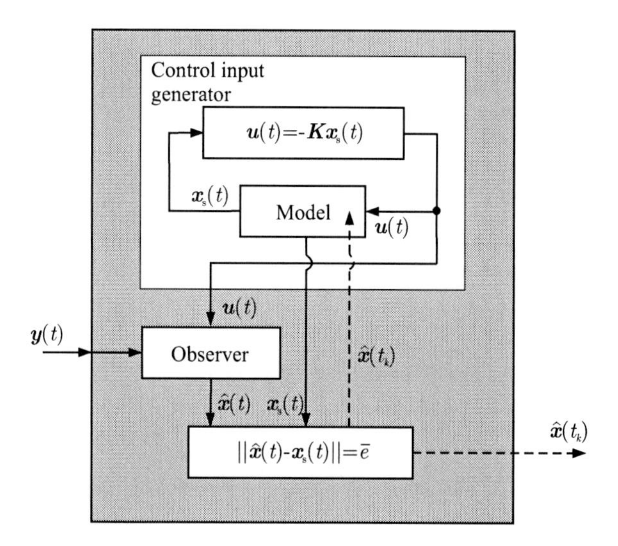

Figure 4.2.: Structure of the extended event generator

Control input generator. The control input generator runs the model

$$\dot{\boldsymbol{x}}_{\mathrm{s}}(t) = \bar{\boldsymbol{A}}\boldsymbol{x}_{\mathrm{s}}(t), \quad \boldsymbol{x}_{\mathrm{s}}(t_k^+) = \hat{\boldsymbol{x}}(t_k) \tag{4.3}$$

$$\boldsymbol{y}_{\mathrm{s}}(t) = \boldsymbol{C}\boldsymbol{x}_{\mathrm{s}}(t) \tag{4.4}$$

$$\boldsymbol{u}(t) = -\boldsymbol{K}\boldsymbol{x}_{\mathrm{s}}(t) \tag{4.5}$$

of the undisturbed continuous-time control loop (3.5), (3.6) to determine the control input $\boldsymbol{u}(t)$ in the time interval $[t_k, t_{k+1})$ between two consecutive events. However, instead of using the state $\boldsymbol{x}(t_k)$, which cannot be measured, the model state $\boldsymbol{x}_{\mathrm{s}}(t)$ is updated with the observer state $\hat{\boldsymbol{x}}(t_k)$, which the control input generator gets from the event generator at event times t_k $(k = 0, 1, 2, ...)$.

Note that the disturbance estimation according to recursion (3.23), (3.24) cannot be applied because the state information $\boldsymbol{x}(t_k)$ is not available. Therefore, the control input generator performs the disturbance estimate $\hat{\boldsymbol{d}}_0 = \boldsymbol{0}$ for all times $t \geq 0$. An approach to circumvent this problem is presented in Section 4.4.2 which uses a disturbance observer to get the disturbance estimate $\hat{\boldsymbol{d}}_k$.

Event generator. Like the control input generator, the event generator runs model (4.3)–(4.5) between two consecutive events. Additionally, the event generator uses the Luenberger state observer

$$\dot{\hat{\boldsymbol{x}}}(t) = \underbrace{(\boldsymbol{A} - \boldsymbol{L}\boldsymbol{C})}_{\bar{\boldsymbol{A}}_{\mathrm{O}}} \hat{\boldsymbol{x}}(t) + \boldsymbol{B}\boldsymbol{u}(t) + \boldsymbol{L}\boldsymbol{y}(t), \quad \hat{\boldsymbol{x}}(0) = \hat{\boldsymbol{x}}_0 \tag{4.6}$$

$$\hat{\boldsymbol{y}}(t) = \boldsymbol{C}\hat{\boldsymbol{x}}(t) \tag{4.7}$$

in order to determine the approximate state $\hat{\boldsymbol{x}}(t)$ of the plant state $\boldsymbol{x}(t)$. Note that due to the disturbance $\boldsymbol{d}(t)$ and the measurement noise $\boldsymbol{v}(t)$, the observer (4.6)–(4.7) is generally not capable of determining the plant state exactly [80]. Thus, $\hat{\boldsymbol{x}}(t) \neq \boldsymbol{x}(t)$ holds for almost all times $t \geq 0$ even in the continuous-time case.

A further important modification compared to the event-based state feedback concerns the event generation. The event condition (3.21) is replaced by the condition

$$\|\hat{\boldsymbol{x}}(t) - \boldsymbol{x}_{\mathrm{s}}(t)\| = \bar{e}_{\mathrm{y}}, \tag{4.8}$$

which uses the available observer state $\hat{\boldsymbol{x}}(t)$ instead of the non-measurable plant state $\boldsymbol{x}(t)$.

At event time t_k, the observer state $\hat{\boldsymbol{x}}(t_k)$ is sent from the event generator towards the control input generator and is used there and in the event generator to update the two copies of model (4.3)–(4.5). Therefore, both copies have an identical behaviour for all times $t \geq 0$.

Remark. A more intuitive choice of the event triggering mechanism is given by considering the difference between the measured output $\boldsymbol{y}(t)$ and the model output $\boldsymbol{y}_{\mathrm{s}}(t)$:

$$\|\boldsymbol{y}(t) - \boldsymbol{y}_{\mathrm{s}}(t)\| = \bar{e}_{\mathrm{y}}.$$

However, this condition does not imply the boundedness of the state difference $\|\boldsymbol{x}(t) - \boldsymbol{x}_{\mathrm{s}}(t)\|$ due to

$$\begin{aligned} \|\boldsymbol{y}(t) - \boldsymbol{y}_{\mathrm{s}}(t)\| &= \|\boldsymbol{C}(\boldsymbol{x}(t) - \boldsymbol{x}_{\mathrm{s}}(t)) + \boldsymbol{v}(t)\| \\ &\leq \|\boldsymbol{C}\| \cdot (\|\boldsymbol{x}(t) - \boldsymbol{x}_{\mathrm{s}}(t)\|) + v_{\max}, \end{aligned}$$

which is essential for the following stability analysis.

Besides, event condition (4.8) becomes intelligible as the observer state $\hat{\boldsymbol{x}}(t)$ depends explicitly on the plant output $\boldsymbol{y}(t)$ according to Eq. (4.6).

4.2.2. Behaviour of the event-based output-feedback loop

With plant (4.1), (4.2), control input generator (4.3)–(4.5) and observer (4.6)–(4.7), the event-based output-feedback loop is described in the time interval $[t_k, t_{k+1})$ by the state-space model

$$\begin{aligned}
\begin{pmatrix} \dot{\boldsymbol{x}}(t) \\ \dot{\boldsymbol{x}}_{\mathrm{s}}(t) \\ \dot{\hat{\boldsymbol{x}}}(t) \end{pmatrix} &= \begin{pmatrix} \boldsymbol{A} & -\boldsymbol{BK} & \boldsymbol{O} \\ \boldsymbol{O} & \bar{\boldsymbol{A}} & \boldsymbol{O} \\ \boldsymbol{LC} & -\boldsymbol{BK} & \bar{\boldsymbol{A}}_{\mathrm{O}} \end{pmatrix} \begin{pmatrix} \boldsymbol{x}(t) \\ \boldsymbol{x}_{\mathrm{s}}(t) \\ \hat{\boldsymbol{x}}(t) \end{pmatrix} + \begin{pmatrix} \boldsymbol{E} \\ \boldsymbol{O} \\ \boldsymbol{O} \end{pmatrix} \boldsymbol{d}(t) + \begin{pmatrix} \boldsymbol{O} \\ \boldsymbol{O} \\ \boldsymbol{L} \end{pmatrix} \boldsymbol{v}(t) \\
\begin{pmatrix} \boldsymbol{x}(t_k) \\ \boldsymbol{x}_{\mathrm{s}}(t_k^+) \\ \hat{\boldsymbol{x}}(t_k) \end{pmatrix} &= \begin{pmatrix} \boldsymbol{x}(t_k) \\ \hat{\boldsymbol{x}}(t_k) \\ \hat{\boldsymbol{x}}(t_k) \end{pmatrix} \\
\boldsymbol{y}(t) &= (\boldsymbol{C} \;\; \boldsymbol{O} \;\; \boldsymbol{O}) \begin{pmatrix} \boldsymbol{x}(t) \\ \boldsymbol{x}_{\mathrm{s}}(t) \\ \hat{\boldsymbol{x}}(t) \end{pmatrix} + \boldsymbol{v}(t).
\end{aligned}$$

The state transformation

$$\begin{pmatrix} \boldsymbol{x}_{\Delta}(t) \\ \boldsymbol{x}_{\mathrm{s}}(t) \\ \hat{\boldsymbol{x}}_{\Delta}(t) \end{pmatrix} = \begin{pmatrix} \boldsymbol{I} & -\boldsymbol{I} & \boldsymbol{O} \\ \boldsymbol{O} & \boldsymbol{I} & \boldsymbol{O} \\ \boldsymbol{I} & \boldsymbol{O} & -\boldsymbol{I} \end{pmatrix} \begin{pmatrix} \boldsymbol{x}(t) \\ \boldsymbol{x}_{\mathrm{s}}(t) \\ \hat{\boldsymbol{x}}(t) \end{pmatrix}, \tag{4.9}$$

which introduces the difference state $\boldsymbol{x}_{\Delta}(t) = \boldsymbol{x}(t) - \boldsymbol{x}_{\mathrm{s}}(t)$ and the observation error

$$\hat{\boldsymbol{x}}_{\Delta}(t) = \boldsymbol{x}(t) - \hat{\boldsymbol{x}}(t),$$

yields the transformed state-space representation

$$\begin{aligned}
\begin{pmatrix} \dot{\boldsymbol{x}}_{\Delta}(t) \\ \dot{\boldsymbol{x}}_{\mathrm{s}}(t) \\ \dot{\hat{\boldsymbol{x}}}_{\Delta}(t) \end{pmatrix} &= \begin{pmatrix} \boldsymbol{A} & \boldsymbol{O} & \boldsymbol{O} \\ \boldsymbol{O} & \bar{\boldsymbol{A}} & \boldsymbol{O} \\ \boldsymbol{O} & \boldsymbol{O} & \bar{\boldsymbol{A}}_{\mathrm{O}} \end{pmatrix} \begin{pmatrix} \boldsymbol{x}_{\Delta}(t) \\ \boldsymbol{x}_{\mathrm{s}}(t) \\ \hat{\boldsymbol{x}}_{\Delta}(t) \end{pmatrix} + \begin{pmatrix} \boldsymbol{E} \\ \boldsymbol{O} \\ \boldsymbol{E} \end{pmatrix} \boldsymbol{d}(t) - \begin{pmatrix} \boldsymbol{O} \\ \boldsymbol{O} \\ \boldsymbol{L} \end{pmatrix} \boldsymbol{v}(t) && (4.10) \\
\begin{pmatrix} \boldsymbol{x}_{\Delta}(t_k^+) \\ \boldsymbol{x}_{\mathrm{s}}(t_k^+) \\ \hat{\boldsymbol{x}}_{\Delta}(t_k) \end{pmatrix} &= \begin{pmatrix} \boldsymbol{x}(t_k) - \hat{\boldsymbol{x}}(t_k) \\ \hat{\boldsymbol{x}}(t_k) \\ \boldsymbol{x}(t_k) - \hat{\boldsymbol{x}}(t_k) \end{pmatrix} && (4.11) \\
\boldsymbol{y}(t) &= (\boldsymbol{C} \;\; \boldsymbol{C} \;\; \boldsymbol{O}) \begin{pmatrix} \boldsymbol{x}_{\Delta}(t) \\ \boldsymbol{x}_{\mathrm{s}}(t) \\ \hat{\boldsymbol{x}}_{\Delta}(t) \end{pmatrix} + \boldsymbol{v}(t).
\end{aligned}$$

The last equation shows that the output $\boldsymbol{y}(t)$ consists of two components (cf. Section 3.3.2)

$$\boldsymbol{y}(t) = \boldsymbol{y}_{\Delta}(t) + \boldsymbol{y}_{\mathrm{s}}(t)$$

with

$$\begin{aligned}
\boldsymbol{y}_{\mathrm{s}}(t) &= \boldsymbol{C}\boldsymbol{x}_{\mathrm{s}}(t) = \boldsymbol{C}\mathrm{e}^{\bar{\boldsymbol{A}}(t-t_k)}\hat{\boldsymbol{x}}(t_k) \\
\boldsymbol{y}_{\Delta}(t) &= \boldsymbol{C}\boldsymbol{x}_{\Delta}(t) + \boldsymbol{v}(t) \\
&= \boldsymbol{C}\mathrm{e}^{\boldsymbol{A}(t-t_k)}\boldsymbol{x}_{\Delta}(t_k^+) + \int_{t_k}^{t} \boldsymbol{C}\mathrm{e}^{\boldsymbol{A}(t-\alpha)}\boldsymbol{E}\boldsymbol{d}(\alpha)\,\mathrm{d}\alpha + \boldsymbol{v}(t)
\end{aligned}$$

and $\boldsymbol{x}_{\Delta}(t_k^+) = \boldsymbol{x}(t_k) - \hat{\boldsymbol{x}}(t_k)$ according to Eq. (4.11). The output $\boldsymbol{y}_{\mathrm{s}}(t)$ describes the behaviour of the undisturbed continuous-time state-feedback loop, whereas the output $\boldsymbol{y}_{\Delta}(t)$ depends on the uncontrolled plant affected by the disturbance $\boldsymbol{d}(t)$, the measurement noise $\boldsymbol{v}(t)$ and the difference state $\boldsymbol{x}_{\Delta}(t_k^+)$ after the update at the event time t_k. Note that the difference state $\boldsymbol{x}_{\Delta}(t_k^+)$ is identical to the observation error $\hat{\boldsymbol{x}}_{\Delta}(t_k)$ at that time instance, i.e. $\hat{\boldsymbol{x}}_{\Delta}(t_k) = \boldsymbol{x}_{\Delta}(t_k^+)$.

A main problem of the subsequent stability analysis results from the fact that the observation error $\hat{\boldsymbol{x}}_{\Delta}(t_k) = \boldsymbol{x}(t_k) - \hat{\boldsymbol{x}}(t_k)$ is not known. Therefore, an upper bound of this uncertainty is firstly derived, which is later used to prove the ultimate boundedness of the event-based output-feedback loop and to specify a lower bound on the minimum inter-event time.

4.2.3. Stability analysis

Boundedness of the observation error. The observer (4.6)–(4.7) continuously gets the measured output $\boldsymbol{y}(t)$ of the plant (4.1), (4.2) as well as the control input $\boldsymbol{u}(t)$ (Fig. 4.2). According to Eq. (4.10), the observation error

$$\hat{\boldsymbol{x}}_{\Delta}(t) = \boldsymbol{x}(t) - \hat{\boldsymbol{x}}(t)$$

between the plant state and the observer state is given by

$$\dot{\hat{\boldsymbol{x}}}_{\Delta}(t) = (\boldsymbol{A} - \boldsymbol{L}\boldsymbol{C})\hat{\boldsymbol{x}}_{\Delta}(t) + \boldsymbol{E}\boldsymbol{d}(t) - \boldsymbol{L}\boldsymbol{v}(t), \quad \hat{\boldsymbol{x}}_{\Delta}(0) = \hat{\boldsymbol{x}}_{\Delta 0}.$$

As the pair $(\boldsymbol{A}, \boldsymbol{C})$ is assumed to be observable, the observer matrix $\boldsymbol{L}$ can be chosen such that the matrix $\bar{\boldsymbol{A}}_{\mathrm{O}} = \boldsymbol{A} - \boldsymbol{L}\boldsymbol{C}$ is Hurwitz [80]. With the bound $d_{\max}$ of the disturbance $\boldsymbol{d}(t)$, the bound $v_{\max}$ of the measurement noise $\boldsymbol{v}(t)$, and a bound $\hat{x}_{\Delta 0,\max}$ on the initial observation error

$$\|\hat{\boldsymbol{x}}_\Delta(0)\| \leq \hat{x}_{\Delta 0,\max},$$

an upper bound of the difference $\hat{\boldsymbol{x}}_\Delta(t)$ can be obtained by means of the relation

$$\begin{aligned} \|\hat{\boldsymbol{x}}_\Delta(t)\| &= \left\| \mathrm{e}^{\bar{\boldsymbol{A}}_\mathrm{O} t}\hat{\boldsymbol{x}}_\Delta(0) + \int_0^t \mathrm{e}^{\bar{\boldsymbol{A}}_\mathrm{O}(t-\alpha)}(\boldsymbol{E}\boldsymbol{d}(\alpha) - \boldsymbol{L}\boldsymbol{v}(\alpha))\,\mathrm{d}\alpha \right\| \\ &\leq \left\| \mathrm{e}^{\bar{\boldsymbol{A}}_\mathrm{O} t}\right\| \hat{x}_{\Delta 0,\max} + \int_0^t \left\| \mathrm{e}^{\bar{\boldsymbol{A}}_\mathrm{O}\alpha}\right\| \,\mathrm{d}\alpha \cdot (\|\boldsymbol{E}\| d_{\max} + \|\boldsymbol{L}\| v_{\max}) \\ &= e_\mathrm{O}(t). \end{aligned} \tag{4.12}$$

This leads to the following result.

Lemma 7. *Assume that*

- *the plant* (4.1), (4.2) *is observable and the matrix* $\bar{\boldsymbol{A}}_\mathrm{O}$ *is Hurwitz, and*
- *the disturbance* $\boldsymbol{d}(t)$*, the measurement noise* $\boldsymbol{v}(t)$ *and the initial observation error* $\hat{\boldsymbol{x}}_\Delta(0)$ *are bounded according to*

$$\begin{aligned} \|\boldsymbol{d}(t)\| &\leq d_{\max} \\ \|\boldsymbol{v}(t)\| &\leq v_{\max} \\ \|\hat{\boldsymbol{x}}_\Delta(0)\| &\leq \hat{x}_{\Delta 0,\max}. \end{aligned}$$

Then, the observation error $\hat{\boldsymbol{x}}_\Delta(t) = \boldsymbol{x}(t) - \hat{\boldsymbol{x}}(t)$ *between the plant state* $\boldsymbol{x}(t)$ *and the observer state* $\hat{\boldsymbol{x}}(t)$ *is bounded from above by*

$$\|\hat{\boldsymbol{x}}_\Delta(t)\| \leq e_\mathrm{O}(t)$$

with $e_\mathrm{O}(t)$ *given by Eq.* (4.12).

Hence, the state $\boldsymbol{x}(t)$ of the event-based output-feedback loop remains, for all times $t \geq 0$, in the bounded surrounding

$$\boldsymbol{x}(t) \in \Omega_\mathrm{O}(\hat{\boldsymbol{x}}(t)) = \{\boldsymbol{x} : \|\boldsymbol{x} - \hat{\boldsymbol{x}}(t)\| \leq e_\mathrm{O}(t)\}$$

of the observer state $\hat{\boldsymbol{x}}(t)$. Considering Eq. (4.12), the bound $e_\mathrm{O}(t)$ is stationarily given by

$$\lim_{t\to\infty} e_\mathrm{O}(t) = \int_0^\infty \left\| \mathrm{e}^{\bar{\boldsymbol{A}}_\mathrm{O}\alpha}\right\| \,\mathrm{d}\alpha \cdot (\|\boldsymbol{E}\| d_{\max} + \|\boldsymbol{L}\| v_{\max}) = e_{\mathrm{O},\infty} \tag{4.13}$$

because the matrix $\bar{A}_O$ is assumed to be Hurwitz. In order to decrease the observation error bound, a faster observer has to be designed. As this increases the norm $\|L\|$, it also increases the influence of the measurement noise, which likewise holds in the continuous-time case [80].

Comparison of the event-based output-feedback loop and the continuous-time state-feedback loop. The event generating mechanism (4.8) ensures that, for all times $t \geq 0$, the difference $x_{s,\Delta}(t) = \hat{x}(t) - x_s(t)$ between the observer state and the model state is bounded according to

$$\|x_{s,\Delta}(t)\| = \|\hat{x}(t) - x_s(t)\| \leq \bar{e}_y.$$

Hence, the difference $x_\Delta(t) = x(t) - x_s(t)$ between the plant state and the model state is bounded as well

$$\begin{aligned}
\|x_\Delta(t)\| &= \|x(t) - x_s(t)\| \\
&= \|x(t) - \hat{x}(t) + \hat{x}(t) - x_s(t)\| \\
&\leq \|x(t) - \hat{x}(t)\| + \|\hat{x}(t) - x_s(t)\| = e_O(t) + \bar{e}_y
\end{aligned}$$

(Eq. (4.8), Lemma 7).

Lemma 8. *The difference $x_\Delta(t) = x(t) - x_s(t)$ between the plant state $x(t)$ and the model state $x_s(t)$ is bounded by*

$$\|x_\Delta(t)\| \leq e_O(t) + \bar{e}_y, \quad \forall t \geq 0$$

with $\bar{e}_y$ the event threshold and $e_O(t)$ given by Eq. (4.12).

In order to prove the ultimate boundedness of the event-based output-feedback loop (4.1)–(4.8), the approximation error $e(t) = x(t) - x_{CT}(t)$ between the state $x(t)$ of the event-based output feedback and the state $x_{CT}(t)$ of the continuous-time state feedback

$$\dot{x}_{CT}(t) = \bar{A}x_{CT}(t) + Ed(t), \quad x_{CT}(0) = x_0 \tag{4.14}$$

$$y_{CT}(t) = Cx_{CT}(t) \tag{4.15}$$

with $\bar{A} = A - BK$ is considered (cf. Section 3.4.1). It holds

$$\begin{aligned}
\dot{e}(t) &= \dot{x}(t) - \dot{x}_{CT}(t) \\
&= Ax(t) - BKx_s(t) + Ed(t) - \bar{A}x_{CT}(t) - Ed(t) \\
&= \bar{A}e(t) + BKx_\Delta(t), \quad e(0) = 0.
\end{aligned}$$

Together with Lemma 8 this gives an upper bound on the difference $\boldsymbol{e}(t)$

$$\|\boldsymbol{e}(t)\| \leq e_{\max,y}(t) = (e_{\mathrm{O}}(t) + \bar{e}_{\mathrm{y}}) \cdot \int_0^\infty \left\| \mathrm{e}^{\bar{\boldsymbol{A}}\alpha} \boldsymbol{B}\boldsymbol{K} \right\| \,\mathrm{d}\alpha, \tag{4.16}$$

which yields the main result of this section.

Theorem 9. *The approximation error* $\boldsymbol{e}(t) = \boldsymbol{x}(t) - \boldsymbol{x}_{\mathrm{CT}}(t)$ *between the state* $\boldsymbol{x}(t)$ *of the event-based output-feedback loop* (4.1)–(4.8) *and the state* $\boldsymbol{x}_{\mathrm{CT}}(t)$ *of the continuous-time state-feedback loop* (4.14), (4.15) *is bounded from above by*

$$\|\boldsymbol{e}(t)\| \leq e_{\max,y}(t)$$

with $e_{\max,y}(t)$ *given by Eq.* (4.16).

The theorem shows that the approximation error is not only brought about by the event-based sampling but also by the observation error. For $e_{\mathrm{O}}(t) = 0$ and $\bar{e}_{\mathrm{y}} = \bar{e}$, the bound $e_{\max,y}(t)$ reduces to the bound $e_{\max}$ derived for the event-based state feedback (Theorem 3, page 42).

4.2.4. Minimum inter-event time

In this section, a lower bound of the minimum time interval $T_{\min}$ between two consecutive event times is derived in order to show that the communication in the event-based output-feedback loop is bounded. For this aim, the difference $\boldsymbol{x}_{\mathrm{s},\Delta}(t) = \hat{\boldsymbol{x}}(t) - \boldsymbol{x}_{\mathrm{s}}(t)$ between the observer state and the model state has to be investigated. According to Eqs. (4.3), (4.6), (4.9), the difference $\boldsymbol{x}_{\mathrm{s},\Delta}(t)$ is described in the time interval $[t_k, t_{k+1})$ by

$$\begin{aligned}
\dot{\boldsymbol{x}}_{\mathrm{s},\Delta}(t) &= \dot{\hat{\boldsymbol{x}}}(t) - \dot{\boldsymbol{x}}_{\mathrm{s}}(t) \\
&= (\boldsymbol{A} - \boldsymbol{L}\boldsymbol{C})\hat{\boldsymbol{x}}(t) - \boldsymbol{B}\boldsymbol{K}\boldsymbol{x}_{\mathrm{s}}(t) + \boldsymbol{L}\boldsymbol{y}(t) - (\boldsymbol{A} - \boldsymbol{B}\boldsymbol{K})\boldsymbol{x}_{\mathrm{s}}(t) \\
&= (\boldsymbol{A} - \boldsymbol{L}\boldsymbol{C})\hat{\boldsymbol{x}}(t) - \boldsymbol{B}\boldsymbol{K}\boldsymbol{x}_{\mathrm{s}}(t) + \boldsymbol{L}\boldsymbol{C}\boldsymbol{x}_{\Delta}(t) + \boldsymbol{L}\boldsymbol{C}\boldsymbol{x}_{\mathrm{s}}(t) + \boldsymbol{L}\boldsymbol{v}(t) \\
&\quad\; - (\boldsymbol{A} - \boldsymbol{B}\boldsymbol{K})\boldsymbol{x}_{\mathrm{s}}(t) \\
&= \bar{\boldsymbol{A}}_{\mathrm{O}}\boldsymbol{x}_{\mathrm{s},\Delta}(t) + \boldsymbol{L}\boldsymbol{C}\boldsymbol{x}_{\Delta}(t) + \boldsymbol{L}\boldsymbol{v}(t) \\
\boldsymbol{x}_{\mathrm{s},\Delta}(t_k^+) &= \boldsymbol{0}, \quad k = 0, 1, 2, \ldots
\end{aligned}$$

which yields

$$\boldsymbol{x}_{\mathrm{s},\Delta}(t) = \int_{t_k}^{t} \mathrm{e}^{\bar{\boldsymbol{A}}_{\mathrm{O}}(t-\alpha)} \boldsymbol{L}(\boldsymbol{C}\boldsymbol{x}_{\Delta}(\alpha) + \boldsymbol{v}(\alpha))\, \mathrm{d}\alpha.$$

As an event is generated whenever the equation

$$\|\boldsymbol{x}_{\mathrm{s},\Delta}(t)\| \;=\; \left\| \int_{t_k}^{t} \mathrm{e}^{\bar{\boldsymbol{A}}_{\mathrm{O}}(t-\alpha)} \boldsymbol{L}(\boldsymbol{C}\boldsymbol{x}_{\Delta}(\alpha) + \boldsymbol{v}(\alpha))\, \mathrm{d}\alpha \right\| = \bar{e}_{\mathrm{y}}$$

is satisfied, the minimum inter-event time T_{min} is given by

$$T_{\mathrm{min}} = \arg\min_{t} \left\{ \left\| \int_{t_k}^{t} \mathrm{e}^{\bar{\boldsymbol{A}}_{\mathrm{O}}(t-\alpha)} \boldsymbol{L}(\boldsymbol{C}\boldsymbol{x}_{\Delta}(\alpha) + \boldsymbol{v}(\alpha))\, \mathrm{d}\alpha \right\| = \bar{e}_{\mathrm{y}} \right\}.$$

A lower bound $\bar{T}$ of the minimum inter-event time T_{min} can be obtained as the upper integral bound for which the relation

$$\bar{T} = \arg\min_{t} \left\{ \int_{0}^{t} \left\| \mathrm{e}^{\bar{\boldsymbol{A}}_{\mathrm{O}}\alpha} \right\| \; \mathrm{d}\alpha = \frac{\bar{e}_{\mathrm{y}}}{\|\boldsymbol{L}\boldsymbol{C}\|(e_{\mathrm{O}}(t) + \bar{e}_{\mathrm{y}}) + \|\boldsymbol{L}\| v_{\mathrm{max}})} \right\} \tag{4.17}$$

holds.

Theorem 10. *In the event-based output-feedback loop* (4.1)–(4.8)*, the minimum inter-event time T_{min} is bounded from below by $\bar{T}$ given by Eq.* (4.17)*.*

Interestingly, the bound does not directly depend on the disturbance magnitude d_{max} but indirectly through the observation error bound $e_{\mathrm{O}}(t)$. It shows that, as expected, a large observation error leads to a more frequent communication.

Example 7 *Event-based output-feedback control*

In this example a SISO system of two interconnected tanks is considered which has been introduced in [80] and which is depicted in Fig. 4.3.

The state-space model is given by

$$\begin{aligned} \dot{\boldsymbol{x}}(t) &= \begin{pmatrix} -0.1 & 0.1 \\ \frac{1}{12} & -\frac{1}{8} \end{pmatrix} \boldsymbol{x}(t) + \begin{pmatrix} 0.1 \\ 0 \end{pmatrix} u(t) + \begin{pmatrix} 0.1 \\ 0 \end{pmatrix} d(t), \quad \boldsymbol{x}(0) = \boldsymbol{x}_0 \\ y(t) &= \begin{pmatrix} 0 & 0.5 \end{pmatrix} \boldsymbol{x}(t), \end{aligned}$$

where the state variable $x_1(t)$ corresponds to the level of tank T1, $x_2(t)$ corresponds to the level of tank T2, and $y(t)$ is the outflow of tank T2, which is assumed to be proportional to

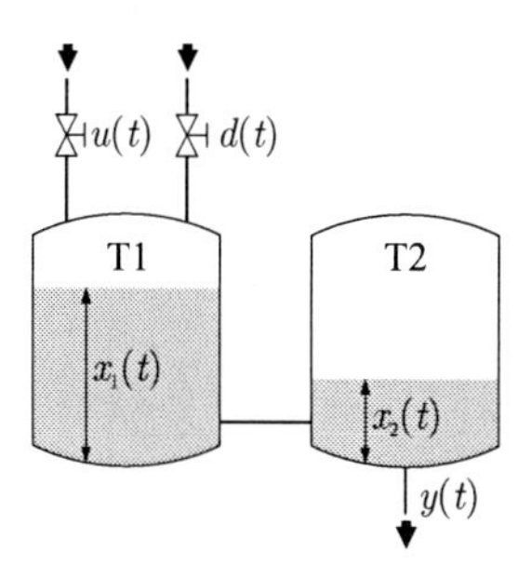

Figure 4.3.: Two interconnected tanks

the level $x_2(t)$. The controller and the observer parameters are

$$\begin{aligned} \boldsymbol{k} &= \begin{pmatrix} 0.64 & 0.41 \end{pmatrix} \\ \boldsymbol{l}' &= \begin{pmatrix} 1.64 & 0.95 \end{pmatrix}. \end{aligned}$$

The event threshold is chosen to be $\bar{e}_{\mathrm{y}} = 1$ and the event condition uses the supremum norm $\|\cdot\|_\infty$. The behaviour of the event-based output-feedback loop affected by the constant

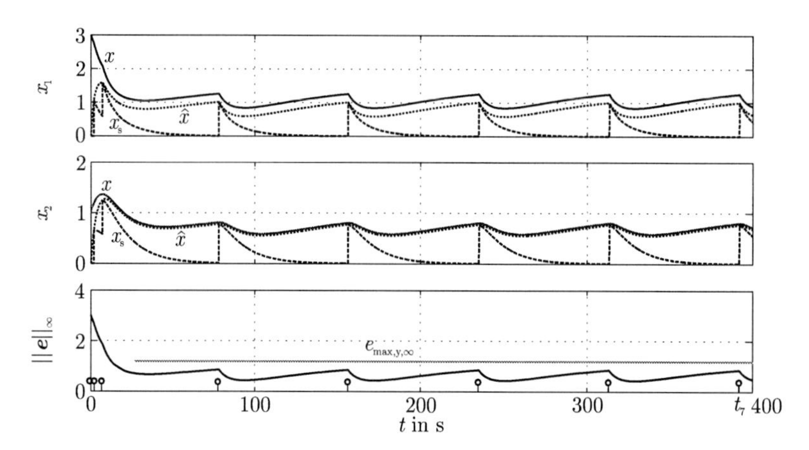

Figure 4.4.: Event-based output-feedback control. Solid lines: plant state $\boldsymbol{x}(t)$; dashed lines: model state $\boldsymbol{x}_{\mathrm{s}}(t)$; dotted lines: observer state $\hat{\boldsymbol{x}}(t)$.

disturbance $d(t) = 0.5 = d_{\max}$ *and the initial observation error*

$$\boldsymbol{x}(0) = \begin{pmatrix} 3 \\ 1 \end{pmatrix} \neq \hat{\boldsymbol{x}}(0) = \boldsymbol{x}_{\mathrm{s}}(0) = \mathbf{0}$$

is depicted in Fig. 4.4 for $v(t) = 0 = v_{\max}$.

After the initialising event at time t_0, *seven further events occur which are invoked by the event condition*

$$|\hat{x}_1(t_k) - x_{\mathrm{s},1}(t_k)| = 1, \quad k = 1, 2, ..., 7.$$

The event times t_k *are indicated in the lower plot. The large initial observation error leads to two consecutive events at the beginning of the simulation. These events move the observer state* $\hat{\boldsymbol{x}}(t)$ *rapidly and compensate the initial uncertainty. Considering relation* (4.17), *this behaviour was to be expected because, through the parameter* $e_{\mathrm{O}}(t)$, *the initial observation error has a strong influence on the communication.*

As the initial observation error is quickly compensated by the observer, relation (4.16) *can be used with the stationary observation error* $e_{\mathrm{O},\infty} = 0.25$ *(Eq.* (4.13)*) to determine the parameter* $e_{\max,\mathrm{y},\infty} = \lim_{t\to\infty} e_{\max,\mathrm{y}}(t)$. *It follows*

$$e_{\max,\mathrm{y},\infty} = 1.16.$$

As shown in the lower plot of Fig. 4.4, this bound is never reached by the actual approximation error $\|\boldsymbol{e}(t)\|_\infty$ *after the transient behaviour at the beginning.*

Note that for the event threshold $\bar{e}_{\mathrm{y}} = \bar{e} = 1$, *the event-based state feedback guarantees a smaller error bound*

$$e_{\max} = 0.93 < e_{\max,\mathrm{y},\infty}$$

according to Eq. (3.32). *To guarantee the same error bound also in the event-based output-feedback case, a smaller event threshold* $\bar{e}_{\mathrm{y}}$ *or a negligible bound on the observation error* $e_{\mathrm{O}}(t)$ *is required. However, in both cases an increased communication is to be expected.*

Applying Eq. (4.17) *with* $e_{\mathrm{O},\infty} = 0.25$ *yields the bound*

$$\bar{T} = 0.8 \text{ s}$$

on the minimum inter-event time, which is suitable at the beginning of the simulation but stationarily very conservative. Therefore, effort should be spent in deriving a bound which can be adapted after the transient behaviour in order to reduce the stationary conservatism.

4.3. Event-based PI control

4.3.1. Basic idea

Section 3.4.5 has shown that the event-based state feedback guarantees setpoint tracking for constant exogenous signals

$$\begin{aligned} \boldsymbol{w}(t) &= \bar{\boldsymbol{w}} \\ \boldsymbol{d}(t) &= \bar{\boldsymbol{d}} \end{aligned} \tag{4.18}$$

only for the undisturbed plant or under very restrictive conditions (Lemma 6, Theorem 6, page 52).

This section extends the event-based state feedback by including a dynamical component into the control input generator and proves that the extended closed-loop system ensures set-point tracking

$$\lim_{t\to\infty} \|\boldsymbol{y}(t) - \bar{\boldsymbol{w}}\| = 0 \tag{4.19}$$

also for the plant subject to constant exogenous signals (4.18).

The basic idea, which has been proposed in [5, 19], results from the well known fact that asymptotic tracking (4.19) occurs in a continuous-time control loop with a linear plant if a controller with integral action is used [79]. This idea is first repeated for the continuous-time state-feedback loop and later transferred to the event-based controller.

Henceforth, the plant is given by

$$\dot{\boldsymbol{x}}(t) = \boldsymbol{A}\boldsymbol{x}(t) + \boldsymbol{B}\boldsymbol{u}(t) + \boldsymbol{E}\boldsymbol{d}(t), \quad \boldsymbol{x}(0) = \boldsymbol{x}_0 \tag{4.20}$$

$$\boldsymbol{y}(t) = \boldsymbol{C}\boldsymbol{x}(t). \tag{4.21}$$

Applying the PI controller

$$\begin{aligned} \boldsymbol{u}(t) &= -\boldsymbol{K}_{\mathrm{P}}\boldsymbol{x}(t) - \boldsymbol{K}_{\mathrm{I}}\boldsymbol{x}_{\mathrm{r}}(t) \\ &= -\boldsymbol{K}\begin{pmatrix} \boldsymbol{x}(t) \\ \boldsymbol{x}_{\mathrm{r}}(t) \end{pmatrix} \end{aligned}$$

with the integrator state $\boldsymbol{x}_{\mathrm{r}} \in \mathbb{R}^r$ given by

$$\dot{\boldsymbol{x}}_{\mathrm{r}}(t) = \boldsymbol{y}(t) - \boldsymbol{w}(t), \qquad \boldsymbol{x}_{\mathrm{r}}(0) = \boldsymbol{x}_{\mathrm{r}0}$$

the continuous-time closed-loop system is represented by the state-space model

$$\begin{pmatrix} \dot{\boldsymbol{x}}_{\mathrm{CT}}(t) \\ \dot{\boldsymbol{x}}_{\mathrm{CTr}}(t) \end{pmatrix} = \underbrace{\begin{pmatrix} \boldsymbol{A} - \boldsymbol{B}\boldsymbol{K}_{\mathrm{P}} & -\boldsymbol{B}\boldsymbol{K}_{\mathrm{I}} \\ \boldsymbol{C} & \boldsymbol{O} \end{pmatrix}}_{\bar{\boldsymbol{A}}_{\mathrm{I}}} \underbrace{\begin{pmatrix} \boldsymbol{x}_{\mathrm{CT}}(t) \\ \boldsymbol{x}_{\mathrm{CTr}}(t) \end{pmatrix}}_{\boldsymbol{x}_{\mathrm{CTI}}(t)} + \underbrace{\begin{pmatrix} \boldsymbol{O} \\ -\boldsymbol{I}_r \end{pmatrix}}_{\boldsymbol{F}_{\mathrm{I}}} \boldsymbol{w}(t) \quad (4.22)$$

$$+ \underbrace{\begin{pmatrix} \boldsymbol{E} \\ \boldsymbol{O} \end{pmatrix}}_{\boldsymbol{E}_{\mathrm{I}}} \boldsymbol{d}(t), \quad \boldsymbol{x}_{\mathrm{CTI}}(0) = \begin{pmatrix} \boldsymbol{x}_0 \\ \boldsymbol{x}_{\mathrm{r}0} \end{pmatrix}$$

$$\boldsymbol{y}_{\mathrm{CT}}(t) = \underbrace{\begin{pmatrix} \boldsymbol{C} & \boldsymbol{O} \end{pmatrix}}_{\boldsymbol{C}_{\mathrm{I}}} \begin{pmatrix} \boldsymbol{x}_{\mathrm{CT}}(t) \\ \boldsymbol{x}_{\mathrm{CTr}}(t) \end{pmatrix}. \quad (4.23)$$

It is assumed that the controller matrix $\boldsymbol{K} = (\boldsymbol{K}_{\mathrm{P}} \;\; \boldsymbol{K}_{\mathrm{I}})$ is chosen so that the matrices $\bar{\boldsymbol{A}}_{\mathrm{I}}$ and $\boldsymbol{A} - \boldsymbol{B}\boldsymbol{K}_{\mathrm{P}}$ are Hurwitz and, hence, the following theorem holds.

Theorem 11. **[83]** *If the matrix* $\bar{\boldsymbol{A}}_{\mathrm{I}}$ *is Hurwitz, then the continuous-time PI-control loop* (4.22), (4.23) *has the setpoint tracking property* (4.19) *for constant exogenous signals* (4.18). *Moreover, for time-varying but bounded disturbances, it guarantees that the continuous-time PI-control loop is ultimately bounded.*

4.3.2. Description of the components

For the time interval $[t_k, t_{k+1})$, the model (4.22) is used with $\boldsymbol{d}(t) = \hat{\boldsymbol{d}}_k$ as the control input generator

$$\begin{pmatrix} \dot{\boldsymbol{x}}_{\mathrm{s}}(t) \\ \dot{\boldsymbol{x}}_{\mathrm{sr}}(t) \end{pmatrix} = \begin{pmatrix} \boldsymbol{A} - \boldsymbol{B}\boldsymbol{K}_{\mathrm{P}} & -\boldsymbol{B}\boldsymbol{K}_{\mathrm{I}} \\ \boldsymbol{C} & \boldsymbol{O} \end{pmatrix} \underbrace{\begin{pmatrix} \boldsymbol{x}_{\mathrm{s}}(t) \\ \boldsymbol{x}_{\mathrm{sr}}(t) \end{pmatrix}}_{\boldsymbol{x}_{\mathrm{sI}}(t)} + \begin{pmatrix} \boldsymbol{O} \\ -\boldsymbol{I}_r \end{pmatrix} \boldsymbol{w}(t) \quad (4.24)$$

$$+ \begin{pmatrix} \boldsymbol{E} \\ \boldsymbol{O} \end{pmatrix} \hat{\boldsymbol{d}}_k, \quad \begin{pmatrix} \boldsymbol{x}_{\mathrm{s}}(t_k^+) \\ \boldsymbol{x}_{\mathrm{sr}}(t_k) \end{pmatrix} = \begin{pmatrix} \boldsymbol{x}(t_k) \\ \boldsymbol{x}_{\mathrm{sr}}(t_k) \end{pmatrix}$$

$$\boldsymbol{u}(t) = -\boldsymbol{K}_{\mathrm{P}}\boldsymbol{x}_{\mathrm{s}}(t) - \boldsymbol{K}_{\mathrm{I}}\boldsymbol{x}_{\mathrm{sr}}(t). \quad (4.25)$$

Note that at event time t_k only the state $\boldsymbol{x}_{\mathrm{s}}(t)$ is updated, whereas the integrator state $\boldsymbol{x}_{\mathrm{sr}}(t)$ behaves continuously. The extended control input generator is referred to as the *PI-control input generator* (Fig. 4.5) and the extended event-based control scheme as *event-based PI control*. The disturbance estimator (3.23), (3.24) and the event condition (3.21), i.e.

$$\|\boldsymbol{x}(t) - \boldsymbol{x}_{\mathrm{s}}(t)\| = \bar{e}, \tag{4.26}$$

remain the same as described in the Sections 3.3.3 and 3.3.4 for the event-based state feedback. However, the event generator now includes the PI-control input generator.

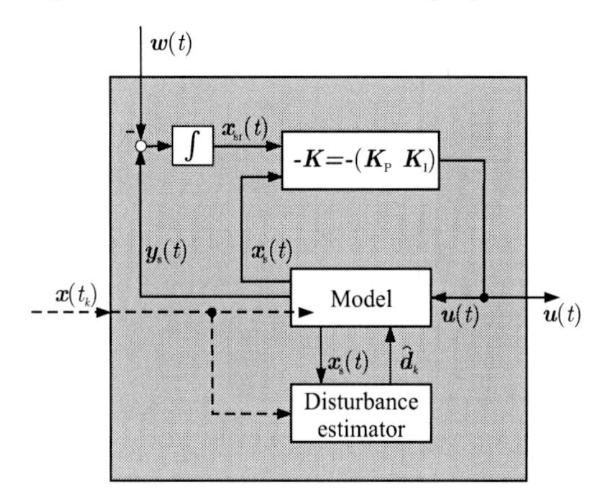

Figure 4.5.: PI-control input generator

Behaviour of the event-based PI-control loop. In the time interval $[t_k, t_{k+1})$ between two consecutive event times, the state behaviour of the event-based PI-control loop (4.20), (4.24), (4.25) can be described by the state-space model

$$\begin{pmatrix} \dot{\boldsymbol{x}}_\Delta(t) \\ \dot{\boldsymbol{x}}_{\mathrm{sI}}(t) \end{pmatrix} = \begin{pmatrix} \boldsymbol{A} & \boldsymbol{O} \\ \boldsymbol{O} & \bar{\boldsymbol{A}}_{\mathrm{I}} \end{pmatrix} \begin{pmatrix} \boldsymbol{x}_\Delta(t) \\ \boldsymbol{x}_{\mathrm{sI}}(t) \end{pmatrix} + \begin{pmatrix} \boldsymbol{O} \\ \boldsymbol{F}_{\mathrm{I}} \end{pmatrix} \boldsymbol{w}(t) + \begin{pmatrix} \boldsymbol{E} \\ \boldsymbol{O} \end{pmatrix} \boldsymbol{d}(t) + \begin{pmatrix} -\boldsymbol{E} \\ \boldsymbol{E}_{\mathrm{I}} \end{pmatrix} \hat{\boldsymbol{d}}_k$$

$$\begin{pmatrix} \boldsymbol{x}_\Delta(t_k^+) \\ \boldsymbol{x}_{\mathrm{sI}}(t_k^+) \end{pmatrix} = \begin{pmatrix} 0 \\ \begin{pmatrix} \boldsymbol{x}(t_k) \\ \boldsymbol{x}_{\mathrm{sr}}(t_k) \end{pmatrix} \end{pmatrix}$$

(see Appendix B.6, page 176). Here, $\boldsymbol{x}_\Delta(t)$ is the difference state $\boldsymbol{x}_\Delta(t) = \boldsymbol{x}(t) - \boldsymbol{x}_{\mathrm{s}}(t)$ for which

$$\boldsymbol{x}_\Delta(t) = \int_{t_k}^{t} \mathrm{e}^{\boldsymbol{A}(t-\alpha)} \boldsymbol{E}(\boldsymbol{d}(\alpha) - \hat{\boldsymbol{d}}_k)\, \mathrm{d}\alpha \tag{4.27}$$

holds. As this equation coincides with Eq. (3.20) derived for the event-based state feedback and, additionally, the event conditions (3.21), (4.26) are identical, the following results can be simply adopted from the state-feedback case (cf. Sections 3.4.3 and 3.4.4).

Proposition 3.

- ***(Theorem 5)*** *For any bounded disturbance, the minimum inter-event time $T_{\min}$ of the event-based PI-control loop is bounded from below by $\bar{T}$ given by*

$$\bar{T} = \arg\min_t \left\{ \int_0^t \left\| \mathrm{e}^{\boldsymbol{A}\alpha} \boldsymbol{E} \right\| \, \mathrm{d}\alpha = \frac{\bar{e}}{\gamma \, d_{\max}} \right\}.$$

- ***(Lemma 5)*** *If the plant* (4.20), (4.21) *is stable, then for every bounded disturbance $\boldsymbol{d}(t) = \bar{d}\tilde{\boldsymbol{d}}(t)$, there exists a magnitude $\bar{d}$ such that the event generator does not generate any event for $t > 0$. The magnitude can be determined according to the relation*

$$|\bar{d}| < \frac{\bar{e}}{\int_0^\infty \|\mathrm{e}^{\boldsymbol{A}\alpha} \boldsymbol{E}\| \, \mathrm{d}\alpha} = \bar{d}_{\mathrm{UD}}. \tag{4.28}$$

4.3.3. Setpoint tracking properties

This section summarises the improved setpoint tracking behaviour of the event-based closed-loop system due to the extension of the control input generator.

Theorem 12. *For constant exogenous signals* (4.18), *the event-based PI-control loop* (3.23), (3.24), (4.20), (4.21), (4.24)–(4.26) *has the property of setpoint tracking* (4.19) *provided that the disturbance is large enough for an event to occur.*

Proof. See Appendix B.6, page 176. □

Discussion of the closed-loop properties. The event-based PI-control loop has the following tracking properties:

- If the disturbance is large enough such that an event occurs at time $t_1 > 0$, then the control input generator determines the right disturbance estimate and the control loop asymptotically reaches the setpoint according to Eq. (4.19).

- If the disturbance is too small for an event to occur at some time $t_1 > 0$, the output $\boldsymbol{y}(t)$ is brought into a surrounding of the setpoint $\bar{\boldsymbol{w}}$. According to Eq. (4.24) the relation $\boldsymbol{y}_{\rm s}(t) = \bar{\boldsymbol{w}}$ holds for large t and the remaining control error depends upon the disturbance (cf. Theorem 6, page 52). As in this situation the disturbance $\bar{\boldsymbol{d}}$ is small enough to satisfy the inequality

$$\|\boldsymbol{x}_\Delta(t)\| = \left\|\boldsymbol{A}^{-1}\left(\mathrm{e}^{\boldsymbol{A}t} - \boldsymbol{I}_n\right)\boldsymbol{E}\bar{\boldsymbol{d}}\right\| < \bar{e}, \quad \forall t \geq 0$$

(Eq. (4.28)), the relation

$$\lim_{t\to\infty} d(\boldsymbol{y}(t), \Omega_{\rm y}(\bar{\boldsymbol{w}})) = 0$$

holds (cf. Eq. 3.45) with

$$\Omega_{\rm y}(\bar{\boldsymbol{w}}) = \{\boldsymbol{y} \,:\, \|\boldsymbol{y} - \bar{\boldsymbol{w}}\| < \|\boldsymbol{C}\|\bar{e}\}. \tag{4.29}$$

Setpoint tracking property for time-varying disturbances. For small time-varying disturbances, a bound on the control error $\boldsymbol{y}(t) - \bar{\boldsymbol{w}}$ is given in the following lemma.

Lemma 9. *Consider a time-varying disturbance $\mathbf{d}(t)$ which satisfies*

$$\left\|\boldsymbol{d}(t) - \hat{\boldsymbol{d}}_k\right\| < \bar{d}_{\rm UD}, \quad \forall t \geq t_k \tag{4.30}$$

with $\bar{d}_{\rm UD}$ defined in Eq. (4.28). Then, no event occurs and the relation

$$\lim_{t\to\infty} d(\boldsymbol{y}(t), \Omega_{\rm y}(\bar{\boldsymbol{w}})) = 0$$

holds with

$$\Omega_{\rm y}(\bar{\boldsymbol{w}}) = \{\boldsymbol{y} \,:\, \|\boldsymbol{y} - \bar{\boldsymbol{w}}\| < \|\boldsymbol{C}\|\bar{e}\}. \tag{4.31}$$

Proof. See Appendix B.7, page 179. □

This result shows that for arbitrary but bounded disturbances $\boldsymbol{d}(t)$ satisfying Eq. (4.30) the output $\boldsymbol{y}(t)$ of the event-based PI-control loop remains stationarily in a bounded region $\Omega_{\rm y}$ around the setpoint $\bar{\boldsymbol{w}}$. As the set $\Omega_{\rm y}$ depends monotonically on the event threshold $\bar{e}$, the deviation of the output from the setpoint can be made arbitrarily small by decreasing the event threshold $\bar{e}$. However, decreasing $\bar{e}$ restricts the range of admissible disturbances $\boldsymbol{d}(t)$ because it is assumed that the disturbance estimation error $\|\boldsymbol{d}(t) - \hat{\boldsymbol{d}}_k\|$ is bounded by $\bar{d}_{\rm UD}$.

4.3.4. Comparison of the event-based PI-control loop and the continuous-time PI-control loop

Theorems 11, 12 and Lemma 9 guarantee that the stationary output difference $\boldsymbol{y}(t) - \boldsymbol{y}_{\mathrm{CT}}(t)$ between the output $\boldsymbol{y}(t)$ of the event-based PI-control loop and the output $\boldsymbol{y}_{\mathrm{CT}}(t)$ of the continuous-time PI-control loop is bounded because both $\boldsymbol{y}(t)$ and $\boldsymbol{y}_{\mathrm{CT}}(t)$ stationarily remain in a bounded surrounding of the setpoint $\bar{\boldsymbol{w}}$.

For constant exogenous signals (4.18), the output difference can be directly determined according to Theorem 12 and Eq. (4.29). Next, this property is derived for the approximation error $\boldsymbol{e}(t) = \boldsymbol{x}(t) - \boldsymbol{x}_{\mathrm{CT}}(t)$.

Lemma 10. *For all disturbances* $\boldsymbol{d}(t)$ *satisfying Eq.* (4.30) *with* $\bar{d}_{\mathrm{UD}}$ *defined in Eq.* (4.28), *there exists a time* $\bar{t} \geq t_k$ *such that the approximation error* $\boldsymbol{e}(t) = \boldsymbol{x}(t) - \boldsymbol{x}_{\mathrm{CT}}(t)$ *between the state* $\boldsymbol{x}(t)$ *of the event-based PI-control loop* (3.23), (3.24), (4.20), (4.21), (4.24)–(4.26) *and the state* $\boldsymbol{x}_{\mathrm{CT}}(t)$ *of the continuous-time PI-control loop* (4.22), (4.23) *is bounded from above by*

$$\|\boldsymbol{e}(t)\| \leq e_{\max,\mathrm{PI}}, \quad \forall t \geq \bar{t}$$

with

$$e_{\max,\mathrm{PI}} = \bar{e} \cdot \int_0^\infty \left\| \mathrm{e}^{(\boldsymbol{A} - \boldsymbol{B}\boldsymbol{K}_{\mathrm{P}})\alpha} \right\| \, \mathrm{d}\alpha \cdot \left(\|\boldsymbol{B}\boldsymbol{K}_{\mathrm{P}}\| + \|\boldsymbol{B}\boldsymbol{K}_{\mathrm{I}}\| x_{\mathrm{srmax},\Delta} \right) \tag{4.32}$$

and

$$\begin{aligned} x_{\mathrm{srmax},\Delta} &= \frac{\int_0^\infty \left\| \mathrm{e}^{\bar{\boldsymbol{A}}_{\mathrm{I}}\alpha} \right\| \, \mathrm{d}\alpha}{\int_0^\infty \left\| \mathrm{e}^{\boldsymbol{A}\alpha} \right\| \, \mathrm{d}\alpha} \\ \bar{\boldsymbol{A}}_{\mathrm{I}} &= \begin{pmatrix} \boldsymbol{A} - \boldsymbol{B}\boldsymbol{K}_{\mathrm{P}} & -\boldsymbol{B}\boldsymbol{K}_{\mathrm{I}} \\ \boldsymbol{C} & \boldsymbol{O} \end{pmatrix}. \end{aligned}$$

Proof. See Appendix B.8, page 179. □

It can be seen that the stationary difference of the state $\boldsymbol{x}(t)$ of the event-based PI-control loop from the state $\boldsymbol{x}_{\mathrm{CT}}(t)$ of the continuous-time closed-loop system can be made arbitrarily small by decreasing the event-threshold $\bar{e}$. Again, decreasing $\bar{e}$ restricts the range of admissible disturbances $\boldsymbol{d}(t)$ according to Eqs. (4.28), (4.30). For $\bar{e} = 0$, the disturbance $\boldsymbol{d}(t)$ has to be constant and exactly known after event time t_k, and the lemma yields

$$\lim_{t\to\infty} \|\boldsymbol{x}(t) - \boldsymbol{x}_{\mathrm{CT}}(t)\| = \lim_{t\to\infty} \|\boldsymbol{y}(t) - \boldsymbol{y}_{\mathrm{CT}}(t)\| = 0.$$

That was to be expected since for a constant disturbance the relation

$$\lim_{t\to\infty} \boldsymbol{y}(t) = \lim_{t\to\infty} \boldsymbol{y}_{\mathrm{CT}}(t) = \bar{\boldsymbol{w}}$$

is obtained according to Theorems 11, 12.

Example 8 *Behaviour of the event-based PI-control loop*

Figure 4.6 shows the same situations as considered in Example 1. However, instead of using the proportional control input generator (3.12), (3.13), the PI-control input generator (4.24), (4.25) with the controller matrix

$$\boldsymbol{K} = \begin{pmatrix} 0.11 & -0.03 & 0.0003 & -0.0003 \\ 0.35 & 1.28 & 0.0009 & 0.0096 \end{pmatrix} \tag{4.33}$$

has been applied. The event condition remains unchanged, i.e.

$$\|\boldsymbol{x}_\Delta(t)\|_\infty = \|\boldsymbol{x}(t) - \boldsymbol{x}_{\mathrm{s}}(t)\|_\infty = 2,$$

and the setpoint $\boldsymbol{w}(t)$ is given by $\boldsymbol{w}(t) = \boldsymbol{0}$.

The behaviour of the event-based PI-control loop for a constant disturbance is depicted in the left subplots of the figure. After the first event at event time t_1, the disturbance magnitude $\bar{d}$ is correctly estimated so that the plant state $\boldsymbol{x}(t)$ and the model state $\boldsymbol{x}_{\mathrm{s}}(t)$ coincide for $t \geq t_1$. Thus, both trajectories stationarily reach the setpoint (Theorem 12):

$$\boldsymbol{x}(t) = \boldsymbol{x}_{\mathrm{s}}(t) = \boldsymbol{0}, \quad \forall t \geq 600\,\mathrm{s}.$$

The middle plots of Fig. 4.6 show the consequence of a small disturbance difference $d(t) - \hat{d}_k$. After event time t_4, the difference $|d(t) - \hat{d}_4|$ satisfies Eq. (4.30) for all future times $t \geq t_4$ so that the states $\boldsymbol{x}(t)$ and $\boldsymbol{x}_{\mathrm{s}}(t)$ diverge. Therefore, only the model state $\boldsymbol{x}_{\mathrm{s}}(t)$ reaches the origin but the state $\boldsymbol{x}(t)$ remains in the set

$$\Omega_{\mathrm{x}}(\boldsymbol{0}) = \{\boldsymbol{x} : \|\boldsymbol{x}\|_\infty < \bar{e}\}$$

(Eq. (4.31), Lemma 9).

Without applying the disturbance estimation both the plant state $\boldsymbol{x}(t)$ and the model state $\boldsymbol{x}_{\mathrm{s}}(t)$ cannot reach the setpoint as events are generated periodically (see right subplots of

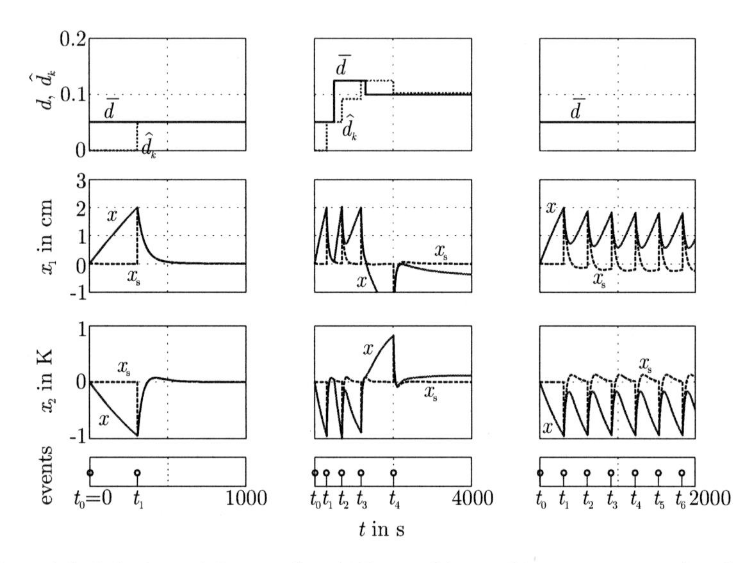

Figure 4.6.: Behaviour of the event-based PI-control loop subject to a constant disturbance (left plots), a time-varying disturbance (middle plots) and without applying the disturbance estimation (right plots). Solid lines: plant state $\boldsymbol{x}(t)$; dashed lines: model state $\boldsymbol{x}_s(t)$.

Fig. 4.6). Note that in all three cases the event times t_k are identical to the event times in Example 1 because the state difference $\boldsymbol{x}_\Delta(t)$ and the event condition are not affected by the PI-extension of the control input generator (Proposition 3).

Further approaches to improve the setpoint tracking property of the event-based state feedback are investigated in [11, 19].

4.4. Improvements of the disturbance estimation

The disturbance estimation (3.23), (3.24) proposed in Section 3.3.4 has two drawbacks:

1. It provides imprecise disturbance estimates in the case of model uncertainties (Section 3.4.6).
2. It requires the existence of the inverse system matrix $\boldsymbol{A}^{-1}$ (Eq. (3.24)).

To overcome the first problem, Section 4.4.1 extends the disturbance estimation by considering model uncertainties as further disturbances. The main aim of this approach is to reduce the communication over the feedback link in the event-based control loop.

The second aspect is faced in Section 4.4.2. The plant model and a disturbance model are combined to an augmented plant model which is used as an extended observer in the event generator (cf. Section 4.2). The observer is able to additionally provide a disturbance estimate at event time t_k.

4.4.1. Augmented disturbance vector

The concept of using an augmented disturbance vector in the disturbance estimator has been studied in [11, 20]. The underlying idea is to treat model uncertainties as further disturbances. This can be obtained by rewriting the state equation of the plant (3.46), (3.47)

$$\begin{aligned}
\dot{\boldsymbol{x}}(t) &= (\boldsymbol{A}+\Delta\boldsymbol{A})\boldsymbol{x}(t) + (\boldsymbol{B}+\Delta\boldsymbol{B})\boldsymbol{u}(t) + (\boldsymbol{E}+\Delta\boldsymbol{E})\boldsymbol{d}(t) \\
&= \boldsymbol{A}\boldsymbol{x}(t) + \boldsymbol{B}\boldsymbol{u}(t) + \underbrace{\begin{pmatrix} \Delta\boldsymbol{A} & \Delta\boldsymbol{B} & \boldsymbol{E} & \Delta\boldsymbol{E} \end{pmatrix}}_{\boldsymbol{E}_{\mathrm{U}}} \underbrace{\begin{pmatrix} \boldsymbol{x}(t) \\ \boldsymbol{u}(t) \\ \boldsymbol{d}(t) \\ \boldsymbol{d}(t) \end{pmatrix}}_{\boldsymbol{d}_{\mathrm{U}}(t)} \\
&= \boldsymbol{A}\boldsymbol{x}(t) + \boldsymbol{B}\boldsymbol{u}(t) + \boldsymbol{E}_{\mathrm{U}}\boldsymbol{d}_{\mathrm{U}}(t), \quad \boldsymbol{x}(0) = \boldsymbol{x}_0
\end{aligned}$$

with $\boldsymbol{E}_{\mathrm{U}} \in \mathbb{R}^{n\times(n+m+l+l)}$, $\boldsymbol{d}_{\mathrm{U}} \in \mathbb{R}^{(n+m+l+l)}$.

By introducing the augmented disturbance vector

$$\boldsymbol{d}_{\mathrm{a}}(t) = \boldsymbol{E}_{\mathrm{U}}\boldsymbol{d}_{\mathrm{U}}(t)$$

with $\boldsymbol{d}_{\mathrm{a}} \in \mathbb{R}^n$, the model

$$\dot{\boldsymbol{x}}(t) = \boldsymbol{A}\boldsymbol{x}(t) + \boldsymbol{B}\boldsymbol{u}(t) + \boldsymbol{d}_{\mathrm{a}}(t), \quad \boldsymbol{x}(0) = \boldsymbol{x}_0$$

results and the control input generator (3.12), (3.13) ($\boldsymbol{w}(t) = \boldsymbol{0}$) and the disturbance estimator (3.23), (3.24) have to be adapted.

Accordingly, the control input generator uses the model

$$\begin{aligned} \dot{\boldsymbol{x}}_{\mathrm{s}}(t) &= \bar{\boldsymbol{A}}\boldsymbol{x}_{\mathrm{s}}(t) + \hat{\boldsymbol{d}}_{\mathrm{a},k}, \qquad \boldsymbol{x}_{\mathrm{s}}(t_k^+) = \boldsymbol{x}(t_k) \\ \boldsymbol{u}(t) &= -\boldsymbol{K}\boldsymbol{x}_{\mathrm{s}}(t) \end{aligned}$$

for generating the control input in the time interval $[t_k, t_k + 1)$, where the disturbance estimate $\hat{\boldsymbol{d}}_{\mathrm{a},k}$ of the augmented disturbance $\boldsymbol{d}_{\mathrm{a}}(t)$ is determined based on the recursion

$$\hat{\boldsymbol{d}}_{\mathrm{a},0} = \boldsymbol{0} \tag{4.34}$$

$$\hat{\boldsymbol{d}}_{\mathrm{a},k} = \hat{\boldsymbol{d}}_{\mathrm{a},k-1} + \left(\boldsymbol{A}^{-1}\left(\mathrm{e}^{\boldsymbol{A}(t_k - t_{k-1})} - \boldsymbol{I}_n\right)\right)^{-1} \cdot \left(\boldsymbol{x}(t_k) - \boldsymbol{x}_{\mathrm{s}}(t_k^-)\right). \tag{4.35}$$

Note that the disturbance estimation according to Eq. (4.35) does not include any pseudoinverse.

Consequently, as usually $n > l$ holds, the augmented disturbance $\boldsymbol{d}_{\mathrm{a}}(t)$ gives the disturbance estimator more (maximum) flexibility to simultaneously determine the effect of model uncertainties and exogenous disturbances. However, in order to guarantee a bounded inter-event time, it has to be proven that the augmented disturbance $\boldsymbol{d}_{\mathrm{a}}(t)$ is bounded.

Proposition 4. *The augmented disturbance $\boldsymbol{d}_{\mathrm{a}}(t)$ is bounded if Theorem 7 holds for a bounded feedback matrix $\boldsymbol{K}$.*

Proof. As by assumption the disturbance $\boldsymbol{d}(t)$ and the model uncertainties are bounded, the plant state $\boldsymbol{x}(t)$ and the control input $\boldsymbol{u}(t)$ have to be bounded to prove the proposition. The state $\boldsymbol{x}(t)$ of the event-based control loop is bounded if the continuous-time state-feedback loop (3.49), (3.50) is stable for any bounded model uncertainties

$$\|\Delta\boldsymbol{A}\| \leq \Delta A_{\max}, \quad \|\Delta\boldsymbol{B}\| \leq \Delta B_{\max}$$

(Theorem 7, page 56). In this case, the input $\boldsymbol{u}(t)$ is also bounded for any bounded feedback $\boldsymbol{K}$ according to $\boldsymbol{u}(t) = -\boldsymbol{K}\boldsymbol{x}_{\mathrm{s}}(t)$ with $\|\boldsymbol{x}(t) - \boldsymbol{x}_{\mathrm{s}}(t)\| \leq \bar{e}$ for all times t. □

Proposition 5. *For any bounded disturbance $\boldsymbol{d}_{\mathrm{a}}(t)$, the minimum inter-event time $T_{\min}$ of the event-based state-feedback loop* (3.12), (3.13), (3.21), (3.46), (3.47), (4.34), (4.35) *is bounded.*

Proof. Since $\boldsymbol{x}(t)$ and $\boldsymbol{d}_{\mathrm{a}}(t)$ are bounded, $\boldsymbol{x}_{\mathrm{s}}(t)$ and $\hat{\boldsymbol{d}}_{\mathrm{a},k}$ are bounded as well and, hence, $\|\boldsymbol{x}_{\Delta}(t)\|$ always satisfies the event condition (3.21) for a time t with $t - t_k > 0$ according to Eq. (3.52) and t_k the previous event time. □

The consequence of this extension on the behaviour of the event-based control loop in the case of model uncertainties is demonstrated in the experimental evaluation in Chapter 6. Moreover, an alternative approach to compensate model uncertainties by implementing a continuous-time adaptive control scheme [33, 67] in an event-based way has been investigated in [16].

4.4.2. Disturbance observation

This section proposes an alternative way to estimate the disturbance $\boldsymbol{d}(t)$ which is based on an extension of the state observer introduced in Section 4.2. Here, the disturbance $\boldsymbol{d}(t)$ is considered as a further state variable which is assumed to be constant

$$\boldsymbol{d}(t) = \bar{\boldsymbol{d}}, \quad \forall t \geq 0.$$

Under this assumption, the plant (4.20), (4.21) can be rewritten as follows:

$$\begin{aligned}
\begin{pmatrix} \dot{\boldsymbol{x}}(t) \\ \dot{\boldsymbol{d}}(t) \end{pmatrix} &= \underbrace{\begin{pmatrix} \boldsymbol{A} & \boldsymbol{E} \\ \boldsymbol{O} & \boldsymbol{O} \end{pmatrix}}_{\boldsymbol{A}_{\mathrm{d}}} \underbrace{\begin{pmatrix} \boldsymbol{x}(t) \\ \boldsymbol{d}(t) \end{pmatrix}}_{\boldsymbol{x}_{\mathrm{d}}(t)} + \begin{pmatrix} \boldsymbol{B} \\ \boldsymbol{O} \end{pmatrix} \boldsymbol{u}(t) \\
\begin{pmatrix} \boldsymbol{x}(0) \\ \boldsymbol{d}(0) \end{pmatrix} &= \begin{pmatrix} \boldsymbol{x}_0 \\ \boldsymbol{d}_0 \end{pmatrix} = \boldsymbol{x}_{\mathrm{d}0} \\
\boldsymbol{y}(t) &= \underbrace{\begin{pmatrix} \boldsymbol{C} & \boldsymbol{O} \end{pmatrix}}_{\boldsymbol{C}_{\mathrm{d}}} \begin{pmatrix} \boldsymbol{x}(t) \\ \boldsymbol{d}(t) \end{pmatrix}.
\end{aligned}$$

If the pair $(\boldsymbol{A}_{\mathrm{d}}, \boldsymbol{C}_{\mathrm{d}})$ is observable, the observer

$$\dot{\hat{\boldsymbol{x}}}_{\mathrm{d}}(t) = (\boldsymbol{A}_{\mathrm{d}} - \boldsymbol{L}\boldsymbol{C}_{\mathrm{d}})\hat{\boldsymbol{x}}_{\mathrm{d}}(t) + \boldsymbol{B}\boldsymbol{u}(t) + \boldsymbol{L}\boldsymbol{y}(t), \quad \hat{\boldsymbol{x}}_{\mathrm{d}}(0) = \hat{\boldsymbol{x}}_{\mathrm{d}0} \tag{4.36}$$

$$\hat{\boldsymbol{d}}(t) = \begin{pmatrix} \boldsymbol{O} & \boldsymbol{I}_l \end{pmatrix} \begin{pmatrix} \hat{\boldsymbol{x}}(t) \\ \hat{\boldsymbol{d}}(t) \end{pmatrix} \tag{4.37}$$

$$\hat{\boldsymbol{x}}(t) = \begin{pmatrix} \boldsymbol{I}_n & \boldsymbol{O} \end{pmatrix} \begin{pmatrix} \hat{\boldsymbol{x}}(t) \\ \hat{\boldsymbol{d}}(t) \end{pmatrix}$$

can be used to provide both a disturbance estimate $\hat{\boldsymbol{d}}(t)$ and an approximate plant state $\hat{\boldsymbol{x}}(t)$ (cf. Section 4.2). Note that at event time t_k, the measured state $\boldsymbol{x}(t_k)$ (event-based state feedback) or the approximate state $\hat{\boldsymbol{x}}(t_k)$ (event-based output feedback) and, additionally, the disturbance estimate $\hat{\boldsymbol{d}}_k = \hat{\boldsymbol{d}}(t_k)$ have to be transmitted from the event generator towards the control input generator. This is illustrated in Fig. 4.7 for the event-based output feedback.

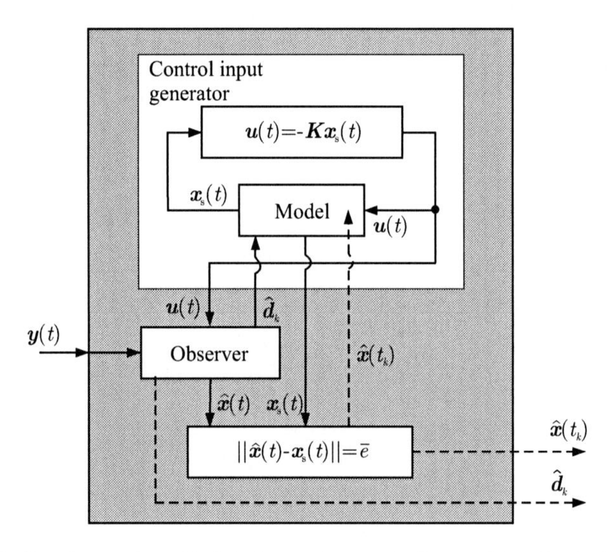

Figure 4.7.: Event generator with disturbance observer for the event-based output feedback

Proposition 6. *For any bounded and constant disturbance* $\boldsymbol{d}(t)$, *the minimum inter-event time* $T_{\min}$ *of the event-based state-feedback loop* (3.12), (3.13), (3.21), (4.20), (4.21), (4.36), (4.37) *is bounded.*

Proof. Since $\boldsymbol{d}(t)$ is assumed to be bounded according to $\|\boldsymbol{d}(t)\| \leq d_{\max}$, $\hat{\boldsymbol{d}}(t)$ is bounded as well because the disturbance observer (4.36), (4.37) guarantees a bounded observation error $\|\hat{\boldsymbol{x}}_{\mathrm{d}\Delta}(t)\| = \|\boldsymbol{x}_{\mathrm{d}}(t) - \hat{\boldsymbol{x}}_{\mathrm{d}}(t)\| \leq d_{\mathrm{O}}$ for constant disturbances with

$$d_{\mathrm{O}} = \max_{t\in[0,\infty)} \left\| \mathrm{e}^{(\boldsymbol{A}_{\mathrm{d}} - \boldsymbol{L}\boldsymbol{C}_{\mathrm{d}})t} \hat{\boldsymbol{x}}_{\mathrm{d}\Delta}(0) \right\|$$

(cf. Section 4.2.3, Lemma 7). Since $\hat{\boldsymbol{d}}(t)$ is bounded, the minimum inter-event time is bounded according to Eq. (3.40), page 48, where $d_{\max}$ has to be replaced by $d_{\max} + d_{\mathrm{O}}$. □

Example 9 *Disturbance estimation by the disturbance observer*

Figure 4.8 shows the behaviour of the event-based state-feedback loop with plant (2.7), (2.8), controller (3.27) and event condition (3.28) subject to the constant disturbance $d(t) = 0.05$ (cf. Example 1, page 37, left subplots of Fig. 3.7). Instead of using the disturbance estimation (3.23), (3.24), the disturbance observer (4.36), (4.37) is used to get the disturbance estimates $\hat{d}_k$.

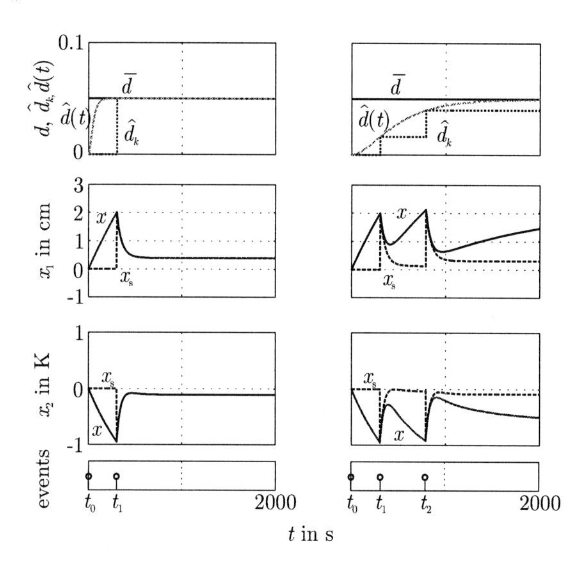

Figure 4.8.: Behaviour of the event-based state-feedback loop subject to a constant disturbance for different disturbance observer gains. Solid lines: plant state $\boldsymbol{x}(t)$; dashed lines: model state $\boldsymbol{x}_{\mathrm{s}}(t)$.

In the left subplots of Fig. 4.8, the observer (4.36), (4.37) has the observer matrix

$$\boldsymbol{L} = \begin{pmatrix} 0.062 & -0.032 \\ -0.012 & 0.056 \\ 0.005 & -0.009 \end{pmatrix}.$$

The upper plot shows that, at event time t_1, the observer is able to provide the correct disturbance estimate $\hat{d}(t_1) = \bar{d}$. The solid line in the plot indicates the actual disturbance $d(t)$ affecting the plant, the grey dashed line indicates the disturbance estimate $\hat{d}(t)$ provided by the disturbance observer and the dotted line indicates the disturbance estimate $\hat{d}_k$ which is transmitted from the event generator and used by the control input generator.

As the disturbance magnitude is exactly estimated, the states $\boldsymbol{x}(t)$ and $\boldsymbol{x}_{\mathrm{s}}(t)$ coincide for all future times $t > t_1$ (cf. Fig. 3.7, page 38).

The right subplots of Fig. 4.8 show the consequences which occur for a slower observer with the observer matrix

$$\boldsymbol{L} = \begin{pmatrix} 0.015 & -0.04 \\ -0.01 & 0.025 \\ 0.001 & -0.001 \end{pmatrix}.$$

As the observer does not converge as fast as in the previous situation (see upper plots of the figure), the disturbance estimate $\hat{d}_1$ distinguishes from the actual disturbance $\bar{d}$ at event time t_1. Thus, the states $\boldsymbol{x}(t)$ and $\boldsymbol{x}_{\mathrm{s}}(t)$ diverge and cause a further event at time t_2. Even though $\hat{d}_2 \neq \bar{d}$ holds for $t \geq t_2$, their difference satisfies inequality (3.42) and no further event takes place.

4.5. Discrete-time implementation of the event-based state feedback

4.5.1. Structure

The event generator and the control input generator have to be implemented on smart components which generally operate in a time-periodic way and, hence, impose a discrete-time sampling (an event-based operating system, i.e. TinyOS, is presented in [75]). The consequences of the discrete-time sampling on the event-based state-feedback loop are [1, 9]:

- A sampler and a zero-order hold (ZOH) have to be included into the event-based control loop (Fig. 4.9).
- No continuous-time control input generation and no continuous-time event generation are possible (Fig. 4.10).

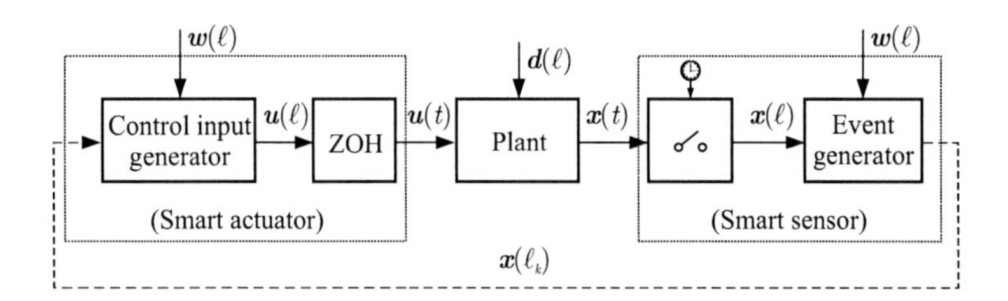

Figure 4.9.: Discrete-time event-based control loop

Throughout this section the discrete-time plant model

$$\boldsymbol{x}(\ell+1) = \boldsymbol{A}_{\rm DT}\boldsymbol{x}(\ell) + \boldsymbol{B}_{\rm DT}\boldsymbol{u}(\ell) + \boldsymbol{E}_{\rm DT}\boldsymbol{d}(\ell), \quad \boldsymbol{x}(0) = \boldsymbol{x}_0 \tag{4.38}$$

$$\boldsymbol{y}(\ell) = \boldsymbol{C}_{\rm DT}\boldsymbol{x}(\ell) \tag{4.39}$$

is considered, which represents the continuous-time plant (4.20), (4.21) together with the zero-order hold and the sampler (cf. Theorem 3.5, page 58). The matrices are given by

$$\boldsymbol{A}_{\rm DT} = {\rm e}^{\boldsymbol{A}T_{\rm s}}, \quad \boldsymbol{B}_{\rm DT} = \int_0^{T_{\rm s}} {\rm e}^{\boldsymbol{A}\alpha}\boldsymbol{B}\,{\rm d}\alpha, \quad \boldsymbol{E}_{\rm DT} = \int_0^{T_{\rm s}} {\rm e}^{\boldsymbol{A}\alpha}\boldsymbol{E}\,{\rm d}\alpha, \quad \boldsymbol{C}_{\rm DT} = \boldsymbol{C}.$$

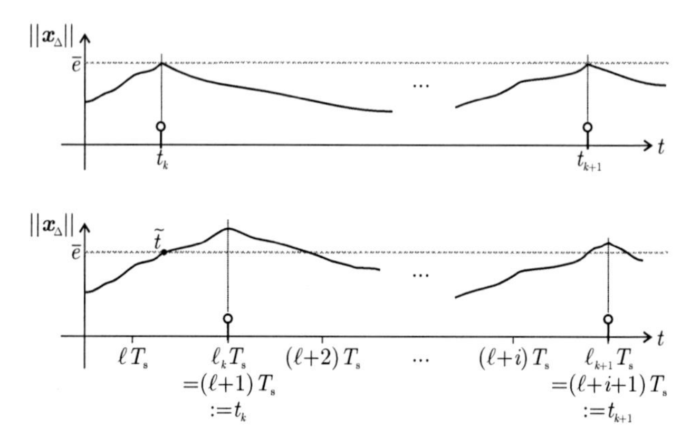

Figure 4.10.: Influence of the discrete-time sampling on the event generation

Figure 4.10 illustrates the influence of the discrete-time sampling on the event generation. In contrast to the continuous-time event detection according to Eq. (3.21) shown in the upper plot of the figure, the event condition cannot precisely maintained at time $\tilde{t}$ and an event is generated at the subsequent sampling time as depicted in the lower plot of the figure. Hence, the discrete-time implementation requires a significant modification of the event generation.

Due to the discrete-time sampling, the event times t_k cannot be precisely determined by the **event generator**. Accordingly, an event is generated if the difference state

$$\boldsymbol{x}_\Delta(\ell) = \boldsymbol{x}(\ell) - \boldsymbol{x}_s(\ell)$$

between the plant state $\boldsymbol{x}(\ell)$ and the state $\boldsymbol{x}_s(\ell)$ of the control input generator (introduced in the next section) has exceeded the event threshold $\bar{e}$ between two consecutive sampling times:

$$\|\boldsymbol{x}_\Delta(\ell-1)\| < \bar{e}, \tag{4.40}$$

$$\|\boldsymbol{x}_\Delta(\ell)\| \geq \bar{e}. \tag{4.41}$$

The sampling time ℓ defines the event instance ℓ_k which corresponds to the event time $t_k = \ell_k T_s$.

Note that in contrast to the discrete-time control loop, which is closed at equidistant time steps, the event-based control loop is closed only at the event times ℓ_k at which the state information $\boldsymbol{x}(\ell_k)$ is sent from the event generator towards the control input generator.

4.5.2. Model

The plant (4.38), (4.39) together with the controller

$$\boldsymbol{u}(\ell) = -\boldsymbol{K}_{\rm DT}\boldsymbol{x}(\ell) + \boldsymbol{V}_{\rm DT}\boldsymbol{w}(\ell)$$

yields the discrete-time state-feedback loop

$$\boldsymbol{x}_{\rm DT}(\ell+1) = \bar{\boldsymbol{A}}_{\rm DT}\boldsymbol{x}_{\rm DT}(\ell) + \boldsymbol{E}_{\rm DT}\boldsymbol{d}(\ell) + \boldsymbol{V}_{\rm DT}\boldsymbol{w}(\ell), \quad \boldsymbol{x}_{\rm DT}(0) = \boldsymbol{x}_0 \tag{4.42}$$

$$\boldsymbol{y}_{\rm DT}(\ell) = \boldsymbol{C}_{\rm DT}\boldsymbol{x}_{\rm DT}(\ell) \tag{4.43}$$

with $\bar{\boldsymbol{A}}_{\rm DT} = \boldsymbol{A}_{\rm DT} - \boldsymbol{B}_{\rm DT}\boldsymbol{K}_{\rm DT}$, where the state-feedback matrix $\boldsymbol{K}_{\rm DT}$ is assumed to be chosen so that the closed-loop system is stable and has satisfactory disturbance attenuation properties. The matrix $\boldsymbol{V}_{\rm DT}$ is chosen so that the discrete-time state feedback has the setpoint tracking property for $\boldsymbol{d}(\ell) = \boldsymbol{0}$ (cf. Section 3.2.1).

The discrete-time state-feedback loop (4.42), (4.43) is used in this section as the reference system whose behaviour should be matched by the discrete-time event-based control loop. Note that in comparison to the continuous-time state feedback, even the discrete-time control loop introduces a degradation of the closed-loop performance because the input $\boldsymbol{u}(\ell)$ is held constant between the consecutive sampling times ℓ and $\ell+1$.

The **control input generator** uses the model (4.42) to generate the control input $\boldsymbol{u}(\ell)$ in the interval $\{\ell_k, ..., \ell_{k+1} - 1\}$ ($[t_k, t_{k+1})$) between two consecutive events according to

$$\boldsymbol{x}_{\rm s}(\ell+1) = \bar{\boldsymbol{A}}_{\rm DT}\boldsymbol{x}_{\rm s}(\ell) + \boldsymbol{E}_{\rm DT}\hat{\boldsymbol{d}}_k + \boldsymbol{V}_{\rm DT}\boldsymbol{w}(\ell), \quad \boldsymbol{x}_{\rm s}(\ell_k^+) = \boldsymbol{x}(\ell_k) \tag{4.44}$$

$$\boldsymbol{u}(\ell) = -\boldsymbol{K}_{\rm DT}\boldsymbol{x}_{\rm s}(\ell) + \boldsymbol{V}_{\rm DT}\boldsymbol{w}(\ell). \tag{4.45}$$

At event time ℓ_k, the model state $\boldsymbol{x}_{\rm s}(\ell_k)$ is updated with the current measurement $\boldsymbol{x}(\ell_k)$. The disturbance estimate $\hat{\boldsymbol{d}}_k$ is determined according to recursion (4.50), (4.51), which is derived next.

4.5.3. Behaviour of the event-based control loop

In the interval $\{\ell_k, ..., \ell_{k+1} - 1\}$ between two consecutive events, the state behaviour of the plant (4.38), (4.39) together with the control input generator (4.44), (4.45) is described by the state-space model

$$\begin{aligned}
\begin{pmatrix} \boldsymbol{x}(\ell+1) \\ \boldsymbol{x}_{\mathrm{s}}(\ell+1) \end{pmatrix} &= \begin{pmatrix} \boldsymbol{A}_{\mathrm{DT}} & -\boldsymbol{B}\boldsymbol{K}_{\mathrm{DT}} \\ \boldsymbol{O} & \bar{\boldsymbol{A}}_{\mathrm{DT}} \end{pmatrix} \begin{pmatrix} \boldsymbol{x}(\ell) \\ \boldsymbol{x}_{\mathrm{s}}(\ell) \end{pmatrix} + \begin{pmatrix} \boldsymbol{E}_{\mathrm{DT}} \\ \boldsymbol{O} \end{pmatrix} \boldsymbol{d}(\ell) \\
&\quad + \begin{pmatrix} \boldsymbol{O} \\ \boldsymbol{E}_{\mathrm{DT}} \end{pmatrix} \hat{\boldsymbol{d}}_k + \begin{pmatrix} \boldsymbol{B}_{\mathrm{DT}}\boldsymbol{V}_{\mathrm{DT}} \\ \boldsymbol{B}_{\mathrm{DT}}\boldsymbol{V}_{\mathrm{DT}} \end{pmatrix} \boldsymbol{w}(\ell) \\
\begin{pmatrix} \boldsymbol{x}(\ell_k) \\ \boldsymbol{x}_{\mathrm{s}}(\ell_k^+) \end{pmatrix} &= \begin{pmatrix} \boldsymbol{x}(\ell_k) \\ \boldsymbol{x}(\ell_k) \end{pmatrix}.
\end{aligned}$$

Using state transformation (3.14), the transformed model

$$\begin{aligned}
\begin{pmatrix} \boldsymbol{x}_\Delta(\ell+1) \\ \boldsymbol{x}_{\mathrm{s}}(\ell+1) \end{pmatrix} &= \begin{pmatrix} \boldsymbol{A}_{\mathrm{DT}} & \boldsymbol{O} \\ \boldsymbol{O} & \bar{\boldsymbol{A}}_{\mathrm{DT}} \end{pmatrix} \begin{pmatrix} \boldsymbol{x}_\Delta(\ell) \\ \boldsymbol{x}_{\mathrm{s}}(\ell) \end{pmatrix} + \begin{pmatrix} \boldsymbol{E}_{\mathrm{DT}} \\ \boldsymbol{O} \end{pmatrix} \boldsymbol{d}(\ell) \\
&\quad + \begin{pmatrix} -\boldsymbol{E}_{\mathrm{DT}} \\ \boldsymbol{E}_{\mathrm{DT}} \end{pmatrix} \hat{\boldsymbol{d}}_k + \begin{pmatrix} \boldsymbol{O} \\ \boldsymbol{B}_{\mathrm{DT}}\boldsymbol{V}_{\mathrm{DT}} \end{pmatrix} \boldsymbol{w}(\ell) \\
\begin{pmatrix} \boldsymbol{x}_\Delta(\ell_k^+) \\ \boldsymbol{x}_{\mathrm{s}}(\ell_k^+) \end{pmatrix} &= \begin{pmatrix} 0 \\ \boldsymbol{x}(\ell_k) \end{pmatrix}
\end{aligned} \tag{4.46}$$

results. The state $\boldsymbol{x}(\ell) = \boldsymbol{x}_{\mathrm{s}}(\ell) + \boldsymbol{x}_\Delta(\ell)$ consists of

$$\boldsymbol{x}_{\mathrm{s}}(\ell) = \bar{\boldsymbol{A}}_{\mathrm{DT}}^{\ell-\ell_k}\boldsymbol{x}(\ell_k) + \sum_{j=\ell_k}^{\ell-1} \bar{\boldsymbol{A}}_{\mathrm{DT}}^{\ell-1-j}\boldsymbol{E}_{\mathrm{DT}}\hat{\boldsymbol{d}}_k + \sum_{j=\ell_k}^{\ell-1} \bar{\boldsymbol{A}}_{\mathrm{DT}}^{\ell-1-j}\boldsymbol{B}_{\mathrm{DT}}\boldsymbol{V}_{\mathrm{DT}}\boldsymbol{w}(j) \tag{4.47}$$

$$\boldsymbol{x}_\Delta(\ell) = \sum_{j=\ell_k}^{\ell-1} \boldsymbol{A}_{\mathrm{DT}}^{\ell-1-j}\boldsymbol{E}_{\mathrm{DT}}(\boldsymbol{d}(j) - \hat{\boldsymbol{d}}_k). \tag{4.48}$$

As in the continuous-time case, the state trajectory $\boldsymbol{x}_{\mathrm{s}}(\ell)$ shows the effect of the constant disturbance $\boldsymbol{d}(\ell) = \hat{\boldsymbol{d}}_k$, which has the same effect in the event-based control loop and in the discrete-time state-feedback loop. The difference state $\boldsymbol{x}_\Delta(\ell)$ describes the difference between the discrete-time state feedback and the event-based control which results from the disturbance estimation error $\boldsymbol{d}(\ell) - \hat{\boldsymbol{d}}_k$ between the actual disturbance and the disturbance estimate.

Disturbance estimator. Analogously to Section 3.3.4, the disturbance estimator is obtained by assuming that the disturbance $\boldsymbol{d}(\ell)$ has been constant between two consecutive event times

$$\boldsymbol{d}(\ell) = \bar{\boldsymbol{d}} \text{ for } \ell \in \{\ell_{k-1}, ..., \ell_k - 1\}$$

and by rearranging the difference trajectory (4.48)

$$\boldsymbol{x}_{\Delta}(\ell) = \boldsymbol{x}(\ell) - \boldsymbol{x}_{\mathrm{s}}(\ell) = \sum_{j=\ell_{k-1}}^{\ell-1} \boldsymbol{A}_{\mathrm{DT}}^{\ell-1-j} \boldsymbol{E}_{\mathrm{DT}} (\boldsymbol{d}(j) - \hat{\boldsymbol{d}}_{k-1}). \tag{4.49}$$

The **disturbance estimation** is carried out as follows:

$$\hat{\boldsymbol{d}}_0 = \boldsymbol{0} \tag{4.50}$$

$$\hat{\boldsymbol{d}}_k = \hat{\boldsymbol{d}}_{k-1} + \left(\sum_{j=\ell_{k-1}}^{\ell_k-1} \boldsymbol{A}_{\mathrm{DT}}^{\ell_k-1-j} \boldsymbol{E}_{\mathrm{DT}} \right)^{+} \cdot \left(\boldsymbol{x}(\ell_k) - \boldsymbol{x}_{\mathrm{s}}(\ell_k^-) \right), \tag{4.51}$$

where $\boldsymbol{x}_{\mathrm{s}}(\ell_k^-)$ is the model state before the update at event time ℓ_k.

Summary of the components. The discrete-time event-based control loop as depicted in Fig. 4.9 consists of

- the plant (4.38), (4.39),
- the control input generator (4.44), (4.45) which also estimates the disturbance by means of Eqs. (4.50), (4.51), and
- the event generator which includes a copy of the control input generator (4.44), (4.45) and the disturbance estimator (4.50), (4.51) and determines the event times ℓ_k according to Eqs. (4.40), (4.41).

At event times ℓ_k $(k = 0, 1, 2, ...)$, the measured state information $\boldsymbol{x}(\ell_k)$ is sent from the event generator towards the control input generator and is used there as well as in the event generator to update the model state $\boldsymbol{x}_{\mathrm{s}}(\ell)$ according to $\boldsymbol{x}_{\mathrm{s}}(\ell_k^+) = \boldsymbol{x}(\ell_k)$ and to determine the new disturbance estimate $\hat{\boldsymbol{d}}_k$. Hence, the models in the control input generator and the event generator work synchronously for all sampling times ℓ.

4.5.4. Stability and communication properties

Comparison of the discrete-time event-based control loop and the discrete-time state-feedback loop.

Lemma 11. *Under the assumption*

$$\|\boldsymbol{d}_\Delta(\ell)\| = \|\boldsymbol{d}(\ell) - \hat{\boldsymbol{d}}_k\| \leq \gamma d_{\max}, \quad \forall \ell, k \geq 0 \tag{4.52}$$

the state $\boldsymbol{x}(\ell)$ remains in the bounded surrounding

$$\Omega_{\mathrm{DT}}(\boldsymbol{x}_{\mathrm{s}}(\ell)) = \{\boldsymbol{x} \, : \, \|\boldsymbol{x} - \boldsymbol{x}_{\mathrm{s}}(\ell)\| < \bar{e} + x_{\mathrm{max,DTEB}}\} \tag{4.53}$$

of the model state $\boldsymbol{x}_{\mathrm{s}}(\ell)$ for all times ℓ with $x_{\mathrm{max,DTEB}}$ given by

$$x_{\mathrm{max,DTEB}} = \|\boldsymbol{A}_{\mathrm{DT}} - \boldsymbol{I}_n\| \cdot \bar{e} + \|\boldsymbol{E}_{\mathrm{DT}}\|\gamma d_{\max}. \tag{4.54}$$

Proof. See Appendix B.9, page 181. □

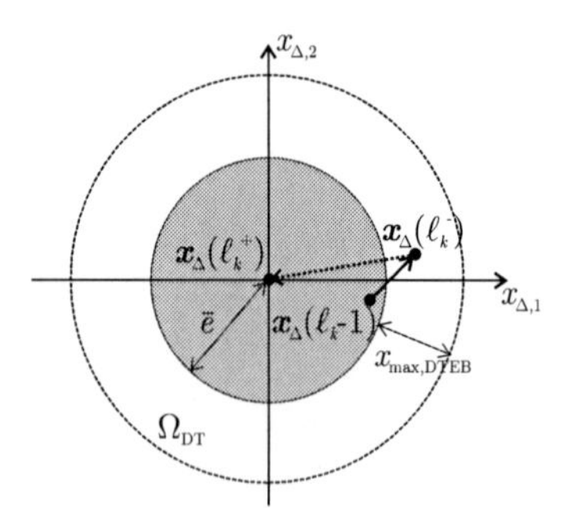

Figure 4.11.: Boundedness of the difference state $\boldsymbol{x}_\Delta(\ell) = \boldsymbol{x}(\ell) - \boldsymbol{x}_{\mathrm{s}}(\ell)$

This lemma, which is illustrated in Fig. 4.11, shows that the state difference $\boldsymbol{x}_\Delta(\ell)$ between the plant state $\boldsymbol{x}(\ell)$ and the model state $\boldsymbol{x}_{\mathrm{s}}(\ell)$ is bounded. It leads to the main result of this section.

Theorem 13. *Assume that* $\|\bar{\boldsymbol{A}}_{\mathrm{DT}}\| < 1$ *holds. Then, the approximation error*

$$\|\boldsymbol{e}(\ell)\| = \|\boldsymbol{x}(\ell) - \boldsymbol{x}_{\mathrm{DT}}(\ell)\|$$

between the state $\boldsymbol{x}(\ell)$ *of the discrete-time event-based control loop* (4.38)–(4.41) (4.44), (4.45), (4.50), (4.51) *and the state* $\boldsymbol{x}_{\mathrm{DT}}(\ell)$ *of the discrete-time state-feedback loop* (4.42), (4.43) *is bounded by*

$$\|\boldsymbol{e}(\ell)\| \leq e_{\mathrm{max,DTEB}} = (\bar{e} + x_{\mathrm{max,DTEB}}) \cdot \frac{\|\boldsymbol{B}_{\mathrm{DT}}\boldsymbol{K}_{\mathrm{DT}}\|}{1 - \|\bar{\boldsymbol{A}}_{\mathrm{DT}}\|} \tag{4.55}$$

with $x_{\mathrm{max,DTEB}}$ *given by Eq.* (4.54).

Proof. See Appendix B.10, page 182. □

Compared to Theorem 3, which considers the continuous-time implementation of event-based control, the main difference is that the bound $e_{\mathrm{max,DTEB}}$ depends not only on the event threshold $\bar{e}$ but additionally on the bound $x_{\mathrm{max,DTEB}}$ which results from the discrete-time sampling. However, for a small sampling period T_{s}, $x_{\mathrm{max,DTEB}} \approx 0$ is obtained because then the relations $\boldsymbol{A}_{\mathrm{DT}} \approx \boldsymbol{I}_n$ and $\boldsymbol{E}_{\mathrm{DT}} \approx \boldsymbol{O}$ hold (see Appendix B.9, page 181) and, hence, the bound $e_{\mathrm{max,DTEB}}$ depends solely on $\bar{e}$ as in the continuous-time case.

Minimum inter-event time. An event is generated whenever the conditions

$$\begin{aligned} \|\boldsymbol{x}_{\Delta}(\ell)\| &= \left\| \sum_{j=\ell_k}^{\ell-1} \boldsymbol{A}_{\mathrm{DT}}^{\ell-1-j} \boldsymbol{E}_{\mathrm{DT}} \boldsymbol{d}_{\Delta}(j) \right\| \geq \bar{e} \\ \|\boldsymbol{x}_{\Delta}(\ell-1)\| &< \bar{e} \end{aligned}$$

are satisfied. The minimum inter-event time ℓ_{min} for which the upper condition is satisfied is given by

$$\ell_{\mathrm{min}} = \arg\min_{\ell} \left\{ \left\| \sum_{j=0}^{\ell-1} \boldsymbol{A}_{\mathrm{DT}}^{\ell-1-j} \boldsymbol{E}_{\mathrm{DT}} \boldsymbol{d}_{\Delta}(j) \right\| \geq \bar{e} \right\}.$$

Assume that inequality (4.52) holds, then a lower bound $\bar{\ell} \leq \ell_{\min}$ of the inter-event time $\ell_{\min}$ can be obtained as follows:

$$\bar{\ell} = \arg\min_{\ell} \left\{ \sum_{j=0}^{\ell-1} \|\boldsymbol{A}_{\mathrm{DT}}^{j} \boldsymbol{E}_{\mathrm{DT}}\| \geq \frac{\bar{e}}{\gamma d_{\max}} \right\}. \tag{4.56}$$

Theorem 14. *For any bounded disturbance, the minimum inter-event time $\ell_{\min}$ of the discrete-time event-based state feedback is bounded from below by $\bar{\ell}$ given by Eq.* (4.56).

Small disturbances. According to event condition (4.40), no event will occur as long as inequality $\|\boldsymbol{x}(\ell) - \boldsymbol{x}_{\mathrm{s}}(\ell)\| < \bar{e}$ holds for all sampling times $\ell \geq 0$. With the disturbance $\boldsymbol{d}(\ell)$ represented by $\boldsymbol{d}(\ell) = \bar{d}\tilde{\boldsymbol{d}}(\ell)$, $\|\tilde{\boldsymbol{d}}(\ell)\| \leq 1$ and $\bar{d}$ the disturbance magnitude (cf. Section 3.4.4), this inequality is true if

$$\|\boldsymbol{x}(\ell) - \boldsymbol{x}_{\mathrm{s}}(\ell)\| = \left\| \sum_{j=0}^{\ell-1} \boldsymbol{A}_{\mathrm{DT}}^{\ell-1-j} \boldsymbol{E}_{\mathrm{DT}} (\bar{d}\boldsymbol{d}(j) - \hat{\boldsymbol{d}}_k) \right\| < \bar{e}$$

and, hence,

$$\max_{\ell \geq 0} \left\| \sum_{j=0}^{\ell-1} \boldsymbol{A}_{\mathrm{DT}}^{\ell-1-j} \boldsymbol{E}_{\mathrm{DT}} (\bar{d}\boldsymbol{d}(j) - \hat{\boldsymbol{d}}_k) \right\| < \bar{e}$$

is satisfied. A bound on the left-hand side of the last inequality exists if the disturbance magnitude $\bar{d}$ is bounded and $\|\boldsymbol{A}_{\mathrm{DT}}\| < 1$ holds.

Under these assumptions and with $\hat{\boldsymbol{d}}_k$ considered to be zero, a bound on $\bar{d}$ can be determined according to

$$\begin{aligned}
\max_{\ell \geq 0} \left\| \sum_{j=\ell_k}^{\ell-1} \boldsymbol{A}_{\mathrm{DT}}^{\ell-1-j} \boldsymbol{E}_{\mathrm{DT}} \bar{d}\tilde{\boldsymbol{d}}(j) \right\| &\leq \sum_{j=0}^{\infty} \|\boldsymbol{A}_{\mathrm{DT}}\|^{j} \cdot \|\boldsymbol{E}_{\mathrm{DT}}\| \cdot \max_{j \geq 0} \|\bar{d}\tilde{\boldsymbol{d}}(j)\| \\
&\leq \sum_{j=0}^{\infty} \|\boldsymbol{A}_{\mathrm{DT}}\|^{j} \cdot \|\boldsymbol{E}_{\mathrm{DT}}\| \cdot |\bar{d}| \\
&= \frac{1}{1 - \|\boldsymbol{A}_{\mathrm{DT}}\|} \cdot \|\boldsymbol{E}_{\mathrm{DT}}\| \cdot |\bar{d}| < \bar{e},
\end{aligned}$$

which leads to

$$|\bar{d}| \; < \; \frac{\bar{e}(1 - \|\boldsymbol{A}_{\mathrm{DT}}\|)}{\|\boldsymbol{E}_{\mathrm{DT}}\|} = \bar{d}_{\mathrm{UDDT}}. \tag{4.57}$$

Lemma 12. *If* $\|\boldsymbol{A}_{\mathrm{DT}}\| < 1$ *holds, then for every bounded disturbance* $\mathbf{d}(\ell) = \bar{d}\tilde{\mathbf{d}}(\ell)$*, there exists a magnitude* $\bar{d}$ *such that the event generator does not generate any event for* $\ell > 0$*. The magnitude can be determined by means of Eq.* (4.57).

Note that the stability of the plant is a necessary condition for this lemma according to the inequality $|\lambda_{\max}\{\boldsymbol{A}_{\mathrm{DT}}\}| \leq \|\boldsymbol{A}_{\mathrm{DT}}\|$.

4.5.5. Discussion

The discrete-time implementation discussed in this section deteriorates the performance of the event-based state feedback with an increasing sampling period T_{s} according to Eqs. (4.53), (4.55). Hence, in order to guarantee the same performance as in the continuous-time case, a smaller event threshold $\bar{e}$ has to be chosen for a nonzero sampling period T_{s}.

However, from an implementation perspective, the discrete-time realisation is more realistic because the smart actuator and the smart sensor usually operate in a time-periodic way. Note that by choosing a relatively small sampling period compared to the time constants of the plant, the effect of the discrete-time sampling can be completely neglected. This can be seen in the experimental evaluation in Chapter 6.

Besides, the discrete-time consideration brings about the following benefit with respect to the computational effort.

> Instead of a demanding continuous-time computation of matrix exponentials as in Eqs. (3.19), (3.20) for the continuous-time scheme, the discrete-time implementation requires only the computation of matrix and vector multiplications and additions at the sampling times ℓ according to Eqs. (4.47), (4.48).

5. Event-based control subject to communication imperfections

The event-based state feedback is investigated in this chapter in situations, where the feedback is realised by a digital communication network. Three communication effects are considered: communication delays (Section 5.2), packet losses (Section 5.3), and data-rate constraints (Section 5.4).

5.1. Consequences of communication imperfections

The load of the digital network used in networked control systems affects the quality of service in terms of communication delays and packet losses, which considerably influence the performance of the control loop and even cause its instability (Section 1.1.2, [62]).

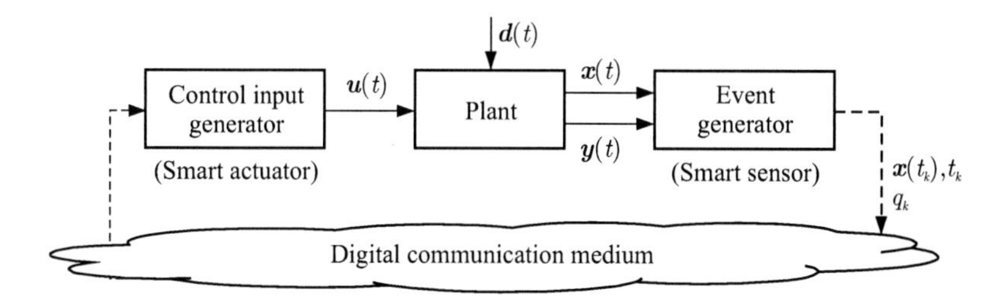

Figure 5.1.: Event-based state-feedback loop closed over a digital communication channel

This chapter analyses the consequences of a non-ideal communication link between the event generator and the control input generator on the performance of the event-based state-feedback loop depicted in Fig. 5.1.

Assumptions. The plant is described by the linear state-space model

$$\dot{\boldsymbol{x}}(t) = \boldsymbol{A}\boldsymbol{x}(t) + \boldsymbol{B}\boldsymbol{u}(t) + \boldsymbol{E}\boldsymbol{d}(t), \qquad \boldsymbol{x}(0) = \boldsymbol{x}_0 \tag{5.1}$$

$$\boldsymbol{y}(t) = \boldsymbol{C}\boldsymbol{x}(t) \tag{5.2}$$

and it is assumed that

- the pair $(\boldsymbol{A},\ \boldsymbol{B})$ is controllable,
- the disturbance $\boldsymbol{d}(t)$ is bounded according to

$$\|\boldsymbol{d}(t)\| \leq d_{\max}, \tag{5.3}$$

- the plant dynamics are accurately known,
- the state $\boldsymbol{x}(t)$ is measurable, and
- the event generator and the control input generator have synchronised clocks.

Moreover, the proportional controller $\boldsymbol{u}(t) = -\boldsymbol{K}\boldsymbol{x}(t)$ is considered throughout this chapter which yields the continuous-time state-feedback loop

$$\dot{\boldsymbol{x}}_{\mathrm{CT}}(t) = \bar{\boldsymbol{A}}\boldsymbol{x}_{\mathrm{CT}}(t) + \boldsymbol{E}\boldsymbol{d}(t), \quad \boldsymbol{x}_{\mathrm{CT}}(0) = \boldsymbol{x}_0 \tag{5.4}$$

$$\boldsymbol{y}_{\mathrm{CT}}(t) = \boldsymbol{C}\boldsymbol{x}_{\mathrm{CT}}(t) \tag{5.5}$$

with $\bar{\boldsymbol{A}} = \boldsymbol{A} - \boldsymbol{B}\boldsymbol{K}$. Again, the continuous-time state-feedback loop (5.4), (5.5) is used as the reference system for evaluating the event-based controller.

However, the previous assumption of ideal communication properties (Section 3.1) cannot be guaranteed by including a digital network. In fact, there are two main consequences:

1. The communication is generally subject to time-varying delays and packet losses, which depend not only on the load of the network but also on specific network properties such as communication protocols, e.g. the *medium access control* (MAC). The influence of communication protocols on the performance of event-based control and discrete-time control has been investigated e.g. in [36, 40, 116]. For a thorough introduction in common communication networks and communication protocols the reader is referred to [115].

2. The network has limited capacities (limited bandwidth) which may impose data-rate constraints. Consequently, only a restricted number of bits can be transmitted over the

network at the sampling times. In order to deal with this situation, quantised information ($q_k \in \{0, 1, 2, ...\}$) have to be sent. The effect of a quantised communication on the performance of a continuous-time and a discrete-time control loop has been studied e.g. in [51, 93]. Note that in the context of NCS, a quantisation likewise results from converting analogue signals into digital signals. This, however, can be usually neglected due to a sufficient resolution of the available quantiser, which is also assumed in this chapter.

Although the main role of event-based control lies in avoiding these communication effects by reducing the load of the network, the consequences cannot be completely circumvented or even result from the event-based character [36, 37].

The event-based state feedback subject to communication delays is investigated in Section 5.2. As the main result, delay bounds are derived which guarantee the ultimate boundedness of the event-based state-feedback loop and allow the specification of an approximation error bound. The results are adapted in Section 5.3 in order to take additional packet losses into account.

Section 5.4 extends the event-based state feedback by incorporating an encoder and a decoder in the control loop. These components allow the transmission and processing of quantised information at event times and ensure a stable behaviour of the event-based control loop with adjustable approximation accuracy and adjustable communication.

5.2. Communication delays

5.2.1. Problem description

Ideal communication. The *event-based state-feedback loop with ideal communication* (Fig. 3.1) discussed in the previous chapters has the following characteristics:

- To generate events and the input signal $\boldsymbol{u}(t)$, the event generator and the control input generator use an identical model of the continuous-time state-feedback loop (Section 3.3.5).
- The states of both models, where $\boldsymbol{x}_{\mathrm{e}}(t)$ is now introduced to denote the model state in the event generator (*state of the event generator*) in order to distinguish it from the state $\boldsymbol{x}_{\mathrm{s}}(t)$ of the control input generator, are synchronously reset at event times t_k $(k = 0, 1, 2, ...)$ with the current plant state $\boldsymbol{x}(t_k)$:

$$\boldsymbol{x}_{\mathrm{s}}(t_k^+) = \boldsymbol{x}_{\mathrm{e}}(t_k^+) = \boldsymbol{x}(t_k).$$

Hence, both states coincide, i.e. $\boldsymbol{x}_{\mathrm{s}}(t) = \boldsymbol{x}_{\mathrm{e}}(t)$ for all times $t \geq 0$ as depicted on the left-hand side of Fig. 5.2.

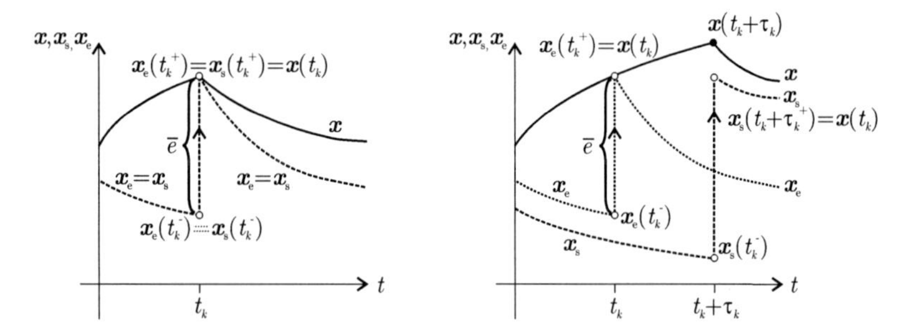

Figure 5.2.: State updating in the event-based state-feedback loop with ideal communication (left) and delayed communication (right). Solid lines: $\boldsymbol{x}(t)$; dashed lines: $\boldsymbol{x}_{\mathrm{s}}(t)$; dotted lines: $\boldsymbol{x}_{\mathrm{e}}(t)$.

Communication delays. Considering delays, which are denoted from now on by τ_k (Fig. 5.1), a synchronous resetting of both models becomes infeasible because the information

$\boldsymbol{x}(t_k)$ sent at event time t_k by the event generator arrives at the control input generator with an unknown delay τ_k at time $t_k + \tau_k$ [6].

Figure 5.2 (right) shows the behaviour of the event-based state-feedback loop affected by a communication delay τ_k. The consequences of the delay are:

- Whereas the event generator immediately knows the new state $\boldsymbol{x}(t_k)$ at time t_k, the control input generator gets this information with the delay τ_k, which is not known for the event generator. Therefore, the two copies of the continuous-time state-feedback loop (3.5), (3.6) used in these components are no longer synchronised with respect to the update of their states. Thus, only the model state $\boldsymbol{x}_{\mathrm{e}}(t_k^+)$ of the event generator is identical to the plant state $\boldsymbol{x}(t_k)$, but the model state $\boldsymbol{x}_{\mathrm{s}}(t)$ of the control input generator generally differs from $\boldsymbol{x}(t)$ for all times t.

- The data transmitted from the event generator towards the control input generator have to be extended to the data package $(\boldsymbol{x}(t_k), t_k)$ (Fig. 5.1). Then, the control input generator is able to determine the delay τ_k.

- As the event generator does not know the delay time τ_k, it does not know the time interval $(t_k, t_k + \tau_k)$ in which the control input generator produces the input signal $\boldsymbol{u}(t)$ with the "old" state information $\boldsymbol{x}(t_{k-1})$ before it updates its state and generates the input according to the new state information $\boldsymbol{x}(t_k)$.

Henceforth, it is assumed that the communication delay τ_k is bounded by the upper bound $\bar{\tau}$

$$\tau_k \leq \bar{\tau}$$

and, in particular, is always smaller than the minimum inter-event time

$$\tau_k < t_{k+1} - t_k \tag{5.6}$$

for all $k \geq 0$.

The analysis will show that, as expected, the performance of the event-based control loop deteriorates with an increasing communication delay τ_k. Nevertheless, it will also show that by appropriately pre-processing the delayed information in the control input generator, a stable behaviour of the event-based state-feedback loop can be guaranteed.

Moreover, the analysis is extended by considering a disturbance estimation in Section 5.2.7 and packet losses in Section 5.3. In summary, the main results are:

- The approximation error between the states of the event-based state-feedback loop and the continuous-time state-feedback loop deteriorates in comparison to the approximation error by the event-based state feedback with ideal communication (Lemma 13).
- By suitably modifying the event generator and the control input generator, the state $\boldsymbol{x}(t)$ of the event-based state-feedback loop remains *GUUB* for bounded delays and an upper bound on the approximation error can be derived (Theorem 15, Lemma 15).
- An upper bound for admissible delays can be determined according to Eqs. (5.22), (5.23) (Lemma 14).
- There exists a minimum inter-event time (Corollary 1).
- By adapting the event condition, similar results also hold in the case of packet losses (Corollary 2, Section 5.3).

5.2.2. Modification of the components

To cope with communication delays, the components of the event-based state-feedback loop are modified as described in [6].

Event generator. In the time interval $[t_k, t_{k+1})$ between two consecutive events, the event generator runs a copy of the model of the undisturbed continuous-time control loop (5.4), (5.5):

$$\dot{\boldsymbol{x}}_{\mathrm{e}}(t) = \bar{\boldsymbol{A}}\boldsymbol{x}_{\mathrm{e}}(t), \quad \boldsymbol{x}_{\mathrm{e}}(t_k^+) = \boldsymbol{x}(t_k). \tag{5.7}$$

Note that first no disturbance estimation is carried out in order to concentrate the analysis on the basic influence of delays. An additional disturbance estimation is considered later in Section 5.2.7. As mentioned before, the state $\boldsymbol{x}_{\mathrm{e}}(t)$ is referred to as the *state of the event generator* in order to distinguish it from the model state $\boldsymbol{x}_{\mathrm{s}}(t)$ of the control input generator.

An event is generated whenever the difference between the plant state $\boldsymbol{x}(t)$ and the state $\boldsymbol{x}_{\mathrm{e}}(t)$ of the event generator reaches the threshold $\bar{e}$. At event time t_k, determined by means of the event condition

$$\|\boldsymbol{x}(t_k) - \boldsymbol{x}_{\mathrm{e}}(t_k^-)\| = \bar{e}, \tag{5.8}$$

the state $\boldsymbol{x}_{\mathrm{e}}(t_k^-)$ is updated with the measured plant state $\boldsymbol{x}(t_k)$, and the state $\boldsymbol{x}(t_k)$ and the event time t_k are sent over the communication link to the control input generator. Here, $\boldsymbol{x}_{\mathrm{e}}(t_k^-)$ denotes the state of the event generator immediately before the update.

The behaviour of model (5.7) between two consecutive events is described by the relation

$$\boldsymbol{x}_{\mathrm{e}}(t) = \mathrm{e}^{\bar{\boldsymbol{A}}(t - t_k)}\boldsymbol{x}(t_k), \quad t_k \leq t < t_{k+1}. \tag{5.9}$$

Control input generator. The control input generator also uses a copy of model (5.7) to produce the control input $\boldsymbol{u}(t)$ in the time interval $[t_k + \tau_k, t_{k+1} + \tau_{k+1})$:

$$\dot{\boldsymbol{x}}_{\mathrm{s}}(t) = \bar{\boldsymbol{A}}\boldsymbol{x}_{\mathrm{s}}(t), \quad \boldsymbol{x}_{\mathrm{s}}(t_k + \tau_k^+) = \boldsymbol{x}_{\mathrm{s}k^+} \tag{5.10}$$

$$\boldsymbol{u}(t) = -\boldsymbol{K}\boldsymbol{x}_{\mathrm{s}}(t). \tag{5.11}$$

Note that the time interval explicitly depends on the communication delays τ_k and τ_{k+1}. The basic question concerns the update of the model state $\boldsymbol{x}_{\mathrm{s}}(t)$ at the delayed event time $t_k + \tau_k$ with appropriate information $\boldsymbol{x}_{\mathrm{s}k^+}$. The next paragraph describes how the update state $\boldsymbol{x}_{\mathrm{s}k^+}$ should be chosen.

Determination of the state $\boldsymbol{x}_{\mathrm{s}k^+}$. The state $\boldsymbol{x}_{\mathrm{s}k^+}$ used for updating the model (5.10), (5.11) in the control input generator at time $t_k + \tau_k$ has a significant influence on the performance of the event-based control loop since it directly affects the control input $\boldsymbol{u}(t)$ for $t \geq t_k + \tau_k$. This section presents an approach to appropriately pre-process the measured information $\boldsymbol{x}(t_k)$ arrived at time $t_k + \tau_k$ at the control input generator. The approach aims at a recovery of the synchronous behaviour of the event generator and the control input generator ($\boldsymbol{x}_{\mathrm{s}}(t) = \boldsymbol{x}_{\mathrm{e}}(t)$) for the time interval $[t_k + \tau_k, t_{k+1})$ as depicted in Fig. 5.3.

The control input generator knows the state $\boldsymbol{x}(t_k)$ and the delay τ_k at time $t_k + \tau_k$ as it gets the time stamp t_k from the event generator. Hence, the relation

$$\boldsymbol{x}_{\mathrm{s}}(t) = \boldsymbol{x}_{\mathrm{e}}(t) \quad \text{for } t \in [t_k + \tau_k, t_{k+1})$$

can be obtained by choosing

$$\boldsymbol{x}_{\mathrm{s}k^+} = \mathrm{e}^{\bar{\boldsymbol{A}}\tau_k}\boldsymbol{x}(t_k) \tag{5.12}$$

because

$$\boldsymbol{x}_{\mathrm{e}}(t_k + \tau_k) = \mathrm{e}^{\bar{\boldsymbol{A}}(t_k + \tau_k - t_k)}\boldsymbol{x}(t_k) = \mathrm{e}^{\bar{\boldsymbol{A}}\tau_k}\boldsymbol{x}(t_k) = \boldsymbol{x}_{\mathrm{s}k^+}$$

holds at time $t_k + \tau_k$ according to Eq. (5.9) and assumption (5.6).

Alternative ways for suitably updating the model state $\boldsymbol{x}_{\mathrm{s}}(t)$ have been proposed in [12, 82].

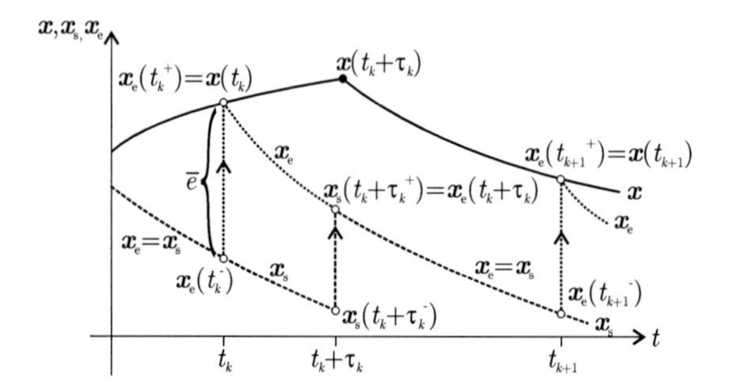

Figure 5.3.: State updating in the event-based state-feedback loop with delayed communication and pre-processing of the transmitted information

Summary of the components. The event-based state-feedback loop as introduced in this section consists of

- the plant (5.1), (5.2),
- the event generator (5.7), (5.8), and
- the control input generator (5.10), (5.11) with the state updating rule (5.12).

At event times t_k $(k = 1, 2, ...)$, the event generator transmits the measured state $\boldsymbol{x}(t_k)$ and the time stamp t_k to the control input generator. The control input generator receives these information with the delay τ_k and the model state $\boldsymbol{x}_{\mathrm{s}}(t_k + \tau_k)$ is updated according to Eq. (5.12).

5.2.3. Behaviour of the event-based control loop

Time interval $t_k + \tau_k \leq t < t_{k+1}$**.** The plant (5.1) and the control input generator (5.10), (5.11) yield the following state model of the closed-loop system:

$$\begin{pmatrix} \dot{\boldsymbol{x}}(t) \\ \dot{\boldsymbol{x}}_{\mathrm{s}}(t) \end{pmatrix} = \begin{pmatrix} \boldsymbol{A} & -\boldsymbol{BK} \\ \boldsymbol{O} & \bar{\boldsymbol{A}} \end{pmatrix} \begin{pmatrix} \boldsymbol{x}(t) \\ \boldsymbol{x}_{\mathrm{s}}(t) \end{pmatrix} + \begin{pmatrix} \boldsymbol{E} \\ \boldsymbol{O} \end{pmatrix} \boldsymbol{d}(t)$$

$$\begin{pmatrix} \boldsymbol{x}(t_k + \tau_k) \\ \boldsymbol{x}_{\mathrm{s}}(t_k + \tau_k^+) \end{pmatrix} = \begin{pmatrix} \boldsymbol{x}(t_k + \tau_k) \\ \boldsymbol{x}_{\mathrm{s}k^+} \end{pmatrix}.$$

For further analysis, the main problem is to determine the state $\boldsymbol{x}(t_k + \tau_k)$, which is not known due to the unknown disturbance $\boldsymbol{d}(t)$. State transformation (3.14) yields the transformed model

$$\begin{aligned}
\begin{pmatrix} \dot{\boldsymbol{x}}_\Delta(t) \\ \dot{\boldsymbol{x}}_\mathrm{s}(t) \end{pmatrix} &= \begin{pmatrix} \boldsymbol{A} & \boldsymbol{O} \\ \boldsymbol{O} & \bar{\boldsymbol{A}} \end{pmatrix} \begin{pmatrix} \boldsymbol{x}_\Delta(t) \\ \boldsymbol{x}_\mathrm{s}(t) \end{pmatrix} + \begin{pmatrix} \boldsymbol{E} \\ \boldsymbol{O} \end{pmatrix} \boldsymbol{d}(t) \\
\begin{pmatrix} \boldsymbol{x}_\Delta(t_k + \tau_k^+) \\ \boldsymbol{x}_\mathrm{s}(t_k + \tau_k^+) \end{pmatrix} &= \begin{pmatrix} \boldsymbol{x}(t_k + \tau_k) - \boldsymbol{x}_{\mathrm{s}k^+} \\ \boldsymbol{x}_{\mathrm{s}k^+} \end{pmatrix},
\end{aligned}$$

where the plant state $\boldsymbol{x}(t)$ is given by $\boldsymbol{x}(t) = \boldsymbol{x}_\mathrm{s}(t) + \boldsymbol{x}_\Delta(t)$ with

$$\begin{aligned}
\boldsymbol{x}_\mathrm{s}(t) &= \mathrm{e}^{\bar{\boldsymbol{A}}(t - t_k - \tau_k)} \boldsymbol{x}_{\mathrm{s}k^+} \\
\boldsymbol{x}_\Delta(t) &= \mathrm{e}^{\boldsymbol{A}(t - t_k - \tau_k)} (\boldsymbol{x}(t_k + \tau_k) - \boldsymbol{x}_{\mathrm{s}k^+}) + \int_{t_k+\tau_k}^{t} \mathrm{e}^{\boldsymbol{A}(t - \alpha)} \boldsymbol{E}\boldsymbol{d}(\alpha)\, \mathrm{d}\alpha.
\end{aligned}$$

In particular, for the state $\boldsymbol{x}_\mathrm{s}(t_{k+1})$ the relation

$$\boldsymbol{x}_\mathrm{s}(t_{k+1}) = \mathrm{e}^{\bar{\boldsymbol{A}}(t_{k+1} - t_k - \tau_k)} \boldsymbol{x}_{\mathrm{s}k^+}$$

is obtained.

Time interval $t_{k+1} \leq t < t_{k+1} + \tau_{k+1}$. The plant (5.1) together with the control input generator (5.10), (5.11) is described by the model

$$\begin{aligned}
\begin{pmatrix} \dot{\boldsymbol{x}}(t) \\ \dot{\boldsymbol{x}}_\mathrm{s}(t) \end{pmatrix} &= \begin{pmatrix} \boldsymbol{A} & -\boldsymbol{B}\boldsymbol{K} \\ \boldsymbol{O} & \bar{\boldsymbol{A}} \end{pmatrix} \begin{pmatrix} \boldsymbol{x}(t) \\ \boldsymbol{x}_\mathrm{s}(t) \end{pmatrix} + \begin{pmatrix} \boldsymbol{E} \\ \boldsymbol{O} \end{pmatrix} \boldsymbol{d}(t) \\
\begin{pmatrix} \boldsymbol{x}(t_{k+1}) \\ \boldsymbol{x}_\mathrm{s}(t_{k+1}) \end{pmatrix} &= \begin{pmatrix} \boldsymbol{x}(t_{k+1}) \\ \boldsymbol{x}_\mathrm{s}(t_{k+1}) \end{pmatrix}.
\end{aligned}$$

Using state transformation (3.14), the equivalent model

$$\begin{aligned}
\begin{pmatrix} \dot{\boldsymbol{x}}_\Delta(t) \\ \dot{\boldsymbol{x}}_\mathrm{s}(t) \end{pmatrix} &= \begin{pmatrix} \boldsymbol{A} & \boldsymbol{O} \\ \boldsymbol{O} & \bar{\boldsymbol{A}} \end{pmatrix} \begin{pmatrix} \boldsymbol{x}_\Delta(t) \\ \boldsymbol{x}_\mathrm{s}(t) \end{pmatrix} + \begin{pmatrix} \boldsymbol{E} \\ \boldsymbol{O} \end{pmatrix} \boldsymbol{d}(t) \\
\begin{pmatrix} \boldsymbol{x}_\Delta(t_{k+1}) \\ \boldsymbol{x}_\mathrm{s}(t_{k+1}) \end{pmatrix} &= \begin{pmatrix} \boldsymbol{x}(t_{k+1}) - \boldsymbol{x}_\mathrm{s}(t_{k+1}) \\ \boldsymbol{x}_\mathrm{s}(t_{k+1}) \end{pmatrix}
\end{aligned}$$

results with

$$\boldsymbol{x}_{\mathrm{s}}(t) = \mathrm{e}^{\bar{\boldsymbol{A}}(t-t_{k+1})}\boldsymbol{x}_{\mathrm{s}}(t_{k+1}) \tag{5.13}$$

$$\boldsymbol{x}_{\Delta}(t) = \mathrm{e}^{\boldsymbol{A}(t-t_{k+1})}(\boldsymbol{x}(t_{k+1}) - \boldsymbol{x}_{\mathrm{s}}(t_{k+1})) + \int_{t_{k+1}}^{t} \mathrm{e}^{\boldsymbol{A}(t-\alpha)}\boldsymbol{E}\boldsymbol{d}(\alpha)\,\mathrm{d}\alpha. \tag{5.14}$$

5.2.4. Comparison of the event-based state-feedback loop and the continuous-time state-feedback loop

The performance of the event-based state feedback with ideal communication properties has been analysed in Section 3.4.1 in comparison with the continuous-time state-feedback loop. It has been proven that the approximation error $\boldsymbol{e}(t)$ between the continuous-time state-feedback loop and the event-based control loop can be made arbitrarily small by accordingly choosing the event threshold $\bar{e}$. Considering communication delays, the following analysis shows that this property of event-based control deteriorates with an increasing communication delay τ_k.

The difference $\boldsymbol{e}(t) = \boldsymbol{x}(t) - \boldsymbol{x}_{\mathrm{CT}}(t)$ between the state $\boldsymbol{x}(t)$ of the event-based state-feedback loop (5.1), (5.2), (5.7), (5.8), (5.10)–(5.12) and the state $\boldsymbol{x}_{\mathrm{CT}}(t)$ of the continuous-time state-feedback loop (5.4), (5.5) is described by the differential equation

$$\begin{aligned}
\dot{\boldsymbol{e}}(t) &= \dot{\boldsymbol{x}}(t) - \dot{\boldsymbol{x}}_{\mathrm{CT}}(t) \\
&= \boldsymbol{A}\boldsymbol{x}(t) - \boldsymbol{B}\boldsymbol{K}\boldsymbol{x}_{\mathrm{s}}(t) + \boldsymbol{E}\boldsymbol{d}(t) - \bar{\boldsymbol{A}}\boldsymbol{x}_{\mathrm{CT}}(t) - \boldsymbol{E}\boldsymbol{d}(t) \\
&= \bar{\boldsymbol{A}}\boldsymbol{e}(t) + \boldsymbol{B}\boldsymbol{K}\boldsymbol{x}_{\Delta}(t), \qquad \boldsymbol{e}(0) = \boldsymbol{0}
\end{aligned}$$

(cf. Eq. (3.31), page 41). An upper bound for the approximation error $\boldsymbol{e}(t)$ is given by

$$\|\boldsymbol{e}(t)\| \leq x_{\Delta\max} \cdot \int_0^{\infty} \left\| \mathrm{e}^{\bar{\boldsymbol{A}}\alpha}\boldsymbol{B}\boldsymbol{K} \right\| \,\mathrm{d}\alpha$$

with

$$x_{\Delta\max} = \max_{t\geq 0} \|\boldsymbol{x}_{\Delta}(t)\|.$$

Determination of $x_{\Delta\max}$. In the delay-free scenario, $\|\boldsymbol{x}_{\Delta}(t)\|$ is bounded by $\|\boldsymbol{x}_{\Delta}(t)\| \leq \bar{e}$ according to Lemma 2. The subsequent analysis yields a bound $x_{\Delta\max}$ for the event-based state-feedback loop with communication delays.

First, the **time interval** $[t_k + \tau_k, t_{k+1})$ is considered. An upper bound of $\|\boldsymbol{x}_{\Delta}(t)\|$ is obtained from the inequality

$$\|\boldsymbol{x}_\Delta(t)\| = \|\boldsymbol{x}(t) - \boldsymbol{x}_{\rm s}(t)\| \leq \|\boldsymbol{x}(t) - \boldsymbol{x}_{\rm e}(t)\| + \|\boldsymbol{x}_{\rm e}(t) - \boldsymbol{x}_{\rm s}(t)\|$$

by deriving upper bounds for the two terms on the right-hand side. As the event generating mechanism resets the state $\boldsymbol{x}_{\rm e}(t)$ at event times t_k immediately with the plant state $\boldsymbol{x}(t_k)$, the first term is bounded by the event threshold $\|\boldsymbol{x}(t) - \boldsymbol{x}_{\rm e}(t)\| \leq \bar{e}$.

Due to the update mechanism (5.12) and assumption (5.6), $\boldsymbol{x}_{\rm e}(t) = \boldsymbol{x}_{\rm s}(t)$ holds for all times t in the time interval considered. Hence, the second term is zero and $\boldsymbol{x}_\Delta(t)$ is bounded by

$$\|\boldsymbol{x}_\Delta(t)\| = \|\boldsymbol{x}(t) - \boldsymbol{x}_{\rm e}(t)\| \leq \bar{e}. \tag{5.15}$$

In the **time interval** $[t_{k+1}, t_{k+1} + \tau_{k+1})$ before the state $\boldsymbol{x}_{\rm s}(t)$ is reset, the difference state $\boldsymbol{x}_\Delta(t)$ according to Eq. (5.14) can be bounded by

$$\begin{aligned} \|\boldsymbol{x}_\Delta(t)\| &\leq \left\|{\rm e}^{\boldsymbol{A}(t - t_{k+1})}\right\| \cdot \|\boldsymbol{x}(t_{k+1}) - \boldsymbol{x}_{\rm s}(t_{k+1})\| + \int_{t_{k+1}}^{t} \left\|{\rm e}^{\boldsymbol{A}(t - \alpha)}\boldsymbol{E}\right\| {\rm d}\alpha \cdot d_{\max} \\ &\leq \max_{\tau \in [0, \bar{\tau}]} \left\|{\rm e}^{\boldsymbol{A}\tau}\right\| \cdot \bar{e} + \int_0^{\bar{\tau}} \left\|{\rm e}^{\boldsymbol{A}\alpha}\boldsymbol{E}\right\| {\rm d}\alpha \cdot d_{\max}, \end{aligned} \tag{5.16}$$

where Eq. (5.15) was used to replace $\|\boldsymbol{x}(t_{k+1}) - \boldsymbol{x}_{\rm s}(t_{k+1})\|$ by the event threshold $\bar{e}$ according to

$$\|\boldsymbol{x}(t_{k+1}) - \boldsymbol{x}_{\rm s}(t_{k+1})\| = \|\boldsymbol{x}(t_{k+1}) - \boldsymbol{x}_{\rm e}(t_{k+1}^-)\| = \bar{e}.$$

In Eq. (5.16), the first term stands for the difference between the measured plant state $\boldsymbol{x}(t_{k+1})$ and the state $\boldsymbol{x}_{\rm s}(t_{k+1})$ of the control input generator and the second term for the effect of the disturbance $\boldsymbol{d}(t)$ in the time interval $[t_{k+1}, t_{k+1} + \tau_{k+1})$.

As the inequality

$$\bar{c}(\bar{\tau}) = \max_{\tau \in [0,\, \bar{\tau}]} \left\|{\rm e}^{\boldsymbol{A}\tau}\right\| \geq 1 \tag{5.17}$$

holds for all $\bar{\tau} \geq 0$, the bound (5.16) is always larger than the bound (5.15) and both inequalities can be summarised as

$$\|\boldsymbol{x}_\Delta(t)\| \leq \bar{c}(\bar{\tau})\, \bar{e} + \bar{d}_{\rm xd}(\bar{\tau})\, d_{\max}, \quad \forall\, t \in [t_k + \tau_k, t_{k+1} + \tau_{k+1})$$

with

$$\bar{d}_{\rm xd}(\bar{\tau}) = \int_0^{\bar{\tau}} \left\|{\rm e}^{\boldsymbol{A}\alpha}\boldsymbol{E}\right\| {\rm d}\alpha. \tag{5.18}$$

Lemma 13. *Consider the event-based state-feedback loop* (5.1), (5.2), (5.7), (5.8), (5.10)–

(5.12) *affected by communication delays. Under the assumption* (5.6), *a bound* $x_{\Delta\max}$ *of the difference* $\|\boldsymbol{x}_{\Delta}(t)\| = \|\boldsymbol{x}(t) - \boldsymbol{x}_{\mathrm{s}}(t)\|$ *is given by*

$$\|\boldsymbol{x}_{\Delta}(t)\| \leq x_{\Delta\max}(\bar{\tau}) = \bar{c}(\bar{\tau})\,\bar{e} + \bar{d}_{\mathrm{xd}}(\bar{\tau})\,d_{\max}, \quad \forall\, t \geq 0 \tag{5.19}$$

with $\bar{c}(\bar{\tau})$ *and* $\bar{d}_{\mathrm{xd}}(\bar{\tau})$ *according to Eqs.* (5.17), (5.18).

5.2.5. Determination of the delay bound

This section derives an upper bound $\tau^{\star}$ on the communication delay τ_k so that assumption (5.6) is satisfied. This bound has to ensure that no event is generated in the time interval $[t_k, t_k + \tau_k)$. In this time interval, the relations

$$\begin{aligned}\boldsymbol{x}_{\mathrm{e}}(t) &= \mathrm{e}^{\bar{\boldsymbol{A}}(t-t_k)}\boldsymbol{x}(t_k)\\ \boldsymbol{x}_{\mathrm{s}}(t) &= \mathrm{e}^{\bar{\boldsymbol{A}}(t-t_k)}\boldsymbol{x}_{\mathrm{s}}(t_k)\end{aligned}$$

hold according to Eqs. (5.9), (5.13), which yield

$$\boldsymbol{x}_{\mathrm{e}}(t) = \boldsymbol{x}_{\mathrm{s}}(t) + \mathrm{e}^{\bar{\boldsymbol{A}}(t-t_k)}(\boldsymbol{x}(t_k) - \boldsymbol{x}_{\mathrm{s}}(t_k)). \tag{5.20}$$

As no event should be generated, the inequality

$$\|\boldsymbol{x}(t) - \boldsymbol{x}_{\mathrm{e}}(t)\| = \left\|\boldsymbol{x}_{\Delta}(t) - \mathrm{e}^{\bar{\boldsymbol{A}}(t-t_k)}(\boldsymbol{x}(t_k) - \boldsymbol{x}_{\mathrm{s}}(t_k))\right\| < \bar{e}$$

has to hold according to Eq. (5.8). Considering this inequality with $\boldsymbol{x}_{\Delta}(t)$ replaced by Eq. (5.14) (t_{k+1} replaced by t_k), then

$$\left\|\left(\mathrm{e}^{\boldsymbol{A}(t-t_k)} - \mathrm{e}^{\bar{\boldsymbol{A}}(t-t_k)}\right)\cdot(\boldsymbol{x}(t_k) - \boldsymbol{x}_{\mathrm{s}}(t_k)) + \int_{t_k}^{t}\mathrm{e}^{\boldsymbol{A}(t-\alpha)}\boldsymbol{E}\boldsymbol{d}(\alpha)\,\mathrm{d}\alpha\right\| < \bar{e}$$

results and an upper bound $\tau^{\star}$ of the delay τ_k can be determined by

$$\tau^{\star} = \min \tau \geq 0 \text{ s.t. } \bar{a}(\tau)\cdot\bar{e} + d_{\mathrm{xd}}(\tau)\,d_{\max} = \bar{e} \tag{5.21}$$

with $\|\boldsymbol{x}(t_k) - \boldsymbol{x}_{\mathrm{s}}(t_k)\| = \bar{e}$ according to Eq. (5.15) and

$$\begin{aligned}\bar{a}(\tau) &= \left\|\mathrm{e}^{\boldsymbol{A}\tau} - \mathrm{e}^{\bar{\boldsymbol{A}}\tau}\right\|\\ d_{\mathrm{xd}}(\tau) &= \int_0^{\tau}\left\|\mathrm{e}^{\boldsymbol{A}\alpha}\boldsymbol{E}\right\|\,\mathrm{d}\alpha.\end{aligned}$$

Equivalently, $\tau^\star$ can be determined by rewriting condition (5.21):

$$\tau^\star = \arg\min_{\tau\in[0,\tilde{\tau}]} \left\{ \frac{d_{\mathrm{xd}}(\tau)\cdot d_{\max}}{1-\bar{a}(\tau)} = \bar{e} \right\} \tag{5.22}$$

$$\tilde{\tau} = \arg\min_{\tau\geq 0} \{\bar{a}(\tau) = 1\}. \tag{5.23}$$

Lemma 14. *For all $\tau \leq \bar{\tau} \leq \tau^\star$, where $\tau^\star$ represents an upper bound for the admissible communication delay obtained by conditions* (5.22) *and* (5.23)*, the assumption* (5.6) *is ensured.*

Since $\tau^\star$ is the maximum delay for which assumption (5.6) is satisfied, it also describes the lower bound $\bar{T} \leq T_{\min}$ of the minimum inter-event time.

Corollary 1. *For any bounded disturbance $\boldsymbol{d}(t)$ and $\tau \leq \bar{\tau} \leq \tau^\star$, the minimum inter-event time $T_{\min}$ of the event-based state-feedback loop is bounded from below by $\bar{T} = \tau^\star$.*

5.2.6. Main result

Lemmas 13 and 14 yield the following bound for the approximation error between the event-based state-feedback loop subject to delays and the continuous-time state-feedback loop with ideal communication.

Theorem 15. *Consider the event-based state-feedback loop* (5.1), (5.2), (5.7), (5.8), (5.10)–(5.12) *affected by communication delay τ. Assume that the maximum delay of the communication network is bounded from above by $\tau \leq \bar{\tau} \leq \tau^\star$, where $\tau^\star$ satisfies the conditions* (5.22), (5.23)*. Then, the approximation error*

$$\boldsymbol{e}(t) = \boldsymbol{x}(t) - \boldsymbol{x}_{\mathrm{CT}}(t)$$

between the state $\boldsymbol{x}(t)$ of the event-based state-feedback loop and the state $\boldsymbol{x}_{\mathrm{CT}}(t)$ of the continuous-time state-feedback loop (5.4), (5.5) *without communication delay is bounded from above by*

$$\|\boldsymbol{e}(t)\| \leq e_{\max,\mathrm{D}}(\bar{\tau})$$

with

$$e_{\max,\mathrm{D}}(\bar{\tau}) = x_{\Delta\max}(\bar{\tau}) \cdot \int_0^\infty \left\| \mathrm{e}^{\bar{\boldsymbol{A}}\alpha} \boldsymbol{B}\boldsymbol{K} \right\| \, \mathrm{d}\alpha \tag{5.24}$$

and $x_{\Delta\max}(\bar{\tau})$ according to Eq. (5.19).

Note that for $\bar{\tau} = 0$ the relation $x_{\Delta\max} = \bar{e}$ holds and, thus, the bound $e_{\max,\mathrm{D}}$ reduces to the bound derived for the event-based state feedback with ideal communication (cf. Theorem 3, page 42). However, for $\bar{\tau} > 0$ and the same event threshold $\bar{e}$, the bound $e_{\max,\mathrm{D}}$ is always larger than the corresponding bound derived for the delay-free scenario.

Moreover, the event-based control loop has the following properties:

- For stable plants, the parameter $\bar{a}$ can be made arbitrarily small by accordingly choosing the controller $\boldsymbol{K}$. Hence, conditions (5.22), (5.23) might be satisfied for arbitrary large delays $\tau \in [0, \infty)$ by accordingly choosing the event threshold $\bar{e}$. This is in contrast to continuous-time control, where a delay bound $\tau^\star$ generally exists [79]. Therefore, as illustrated in the subsequent simulation, the event-based control may outperform continuous-time control in the sense that even if the continuous-time control loop becomes unstable due to the delay, the behaviour of the event-based control loop remains bounded.

- The conditions (5.22) and (5.23) are sufficient conditions for the delay τ for which a stable behaviour of the event-based control loop is guaranteed. Hence, there might exist larger delays $\tau > \tau^\star$ for which the event-based control loop remains bounded.

Example 10 *Behaviour of the event-based state-feedback loop subject to delays*

To emphasise the possible effects of communication delays, the thermofluid process (2.7), (2.8) is slightly modified by limiting the admissible disturbance magnitude and increasing the temperature of the inflow into tank TB (cf. Section 2.2 and Appendix A, page 169). The resulting plant model is given by

$$\begin{aligned}
\dot{\boldsymbol{x}}(t) &= 10^{-3}\begin{pmatrix} -0.8 & 0 \\ -1\cdot 10^{-7} & -1.7 \end{pmatrix}\boldsymbol{x}(t) + 10^{-3}\begin{pmatrix} 211 & 0 \\ 108 & 20 \end{pmatrix}\boldsymbol{u}(t) \qquad (5.25)\\
&\quad + 10^{-3}\begin{pmatrix} 70 \\ -40 \end{pmatrix} d(t)\\
\boldsymbol{x}(0) &= \boldsymbol{x}_{\mathrm{s}}(0) = \boldsymbol{x}_{\mathrm{e}}(0) = \boldsymbol{0}\\
\boldsymbol{y}(t) &= \begin{pmatrix} 1 & 0 \\ 0 & 1 \end{pmatrix}\boldsymbol{x}(t).
\end{aligned}$$

The controller (3.27) and the event condition (3.28) with the event threshold $\bar{e} = 2$ remain unchanged [6].

First, the maximum admissible delays $\bar{\tau}$ are derived which guarantee a stable behaviour for the event-based state-feedback loop and the continuous-time control loop. Analysing the

Nyquist plot of the open-loop transfer matrix [79] with plant (5.25) and controller (3.27) yields the delay bound

$$\bar{\tau}_{\mathrm{CT}} = 75 \text{ s}$$

for the continuous-time state-feedback loop. Thus, the continuous-time state-feedback loop is unstable for all delays $\tau > 75$ s.

For event-based control, the bound

$$\bar{\tau}_{\mathrm{EB}} = \tau^{\star} = 90 \text{ s} \tag{5.26}$$

is obtained by solving conditions (5.22), (5.23) with a standard nonlinear equation solver. It is apparently larger than the admissible delay in the continuous-time case. The reason for this is the following: Instead of continuously using delayed, i.e. wrong, information about the plant state $\boldsymbol{x}(t)$ *for determining the input* $\boldsymbol{u}(t)$ *as it is the case in the continuous-time state-feedback loop, the event-based control loop adapts at time* $t_k + \tau_k$ *the transmitted information* $\boldsymbol{x}(t_k)$ *to the current delay* τ_k *in order to produce a more appropriate input signal* $\boldsymbol{u}(t)$.

However, several methods exist to design controllers or modify the closed-loop structure in the continuous-time case in order to better counteract the influence of delays [108].

Figure 5.4 shows the behaviour of the event-based control loop subject to the constant communication delay $\tau = 77$ s. *In the upper and the second plot the behaviour of the states* $x_1(t)$ *and* $x_2(t)$ *is depicted as well as the behaviour of the corresponding states of the event generator* $\boldsymbol{x}_{\mathrm{e}}(t)$ *and the control input generator* $\boldsymbol{x}_{\mathrm{s}}(t)$. *The event times* t_k, *at which the information* $(\boldsymbol{x}(t_k), t_k)$ *is sent from the event generator towards the control input generator, are marked in the lower plot. After the initialising event at time* t_0, *overall 11 events occur in the time interval considered due to*

$$|x_1(t_k) - x_{\mathrm{e},1}(t_k)| = 2, \quad \forall\, k \geq 1.$$

The figures clearly demonstrate how the update mechanism works. At event times t_k *(*$k = 1, 2, \ldots$*), first only the state* $\boldsymbol{x}_{\mathrm{e}}(t_k)$ *is updated with the measured plant state* $\boldsymbol{x}(t_k)$, *whereas at time* $t_k + \tau_k$ *the state* $\boldsymbol{x}_{\mathrm{s}}(t_k + \tau_k)$ *of the control input generator is set to the current state* $\boldsymbol{x}_{\mathrm{e}}(t_k + \tau_k)$ *of the event generator and the control input* $\boldsymbol{u}(t)$ *is adapted accordingly. This can be seen in the third and fourth subplot of the figure, which show the input signals* $u_1(t)$ *and* $u_2(t)$, *and by the change of the plant behaviour* $\boldsymbol{x}(t)$ *at times* $t_k + \tau_k$.

The states $\boldsymbol{x}_{\mathrm{s}}(t)$ *and* $\boldsymbol{x}_{\mathrm{e}}(t)$ *are identical until the next event time. As expected, the states* $x_1(t)$ *and* $x_2(t)$ *remain bounded since the delay considered is smaller than the bound given in Eq. (5.26). Hence, the simulation shows that the event-based state-feedback loop has a better performance than the continuous-time control loop because the continuous-time control loop is unstable for the delay considered (*$\bar{\tau}_{\mathrm{CT}} < 77$ s*).*

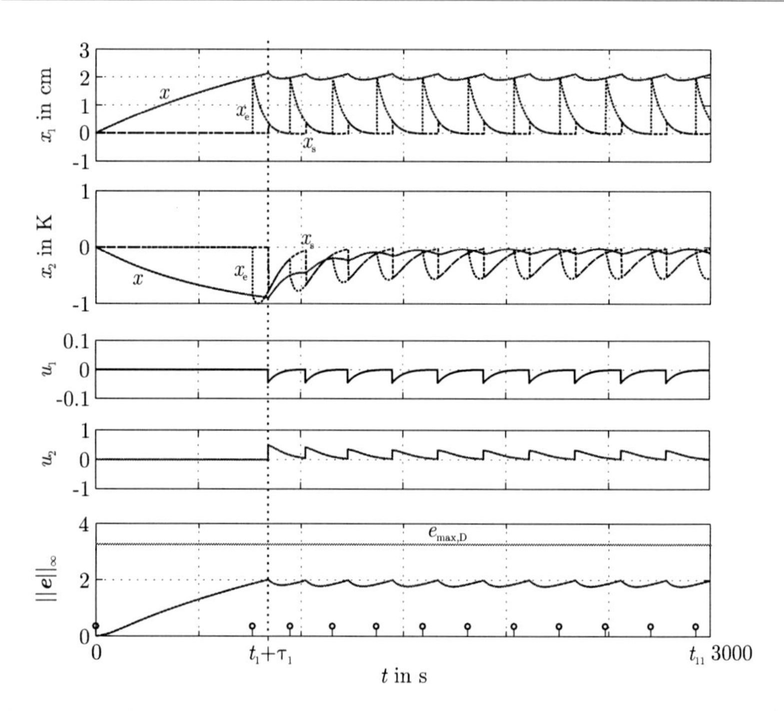

Figure 5.4.: Behaviour of the event-based state-feedback loop with $\tau = 77$ s*. Solid lines:* $\boldsymbol{x}(t)$*; dashed lines:* $\boldsymbol{x}_s(t)$*; dotted lines:* $\boldsymbol{x}_e(t)$*.*

The lower plot of Fig. 5.4 shows the difference $\|\boldsymbol{e}(t)\|_\infty = \|\boldsymbol{x}(t) - \boldsymbol{x}_{\mathrm{CT}}(t)\|_\infty$ between the state $\boldsymbol{x}(t)$ of the event-based control loop and the state $\boldsymbol{x}_{\mathrm{CT}}(t)$ of the undelayed continuous-time state feedback (3.27), (5.25). Applying Eqs. (5.19), (5.24) with $d_{\max} = 0.05$ and $\bar{\tau} = 90$ s yields the upper bound

$$e_{\max,\mathrm{D}} = 3.3$$

for the approximation error bound. It can be seen that the inequality $\|\boldsymbol{e}(t)\|_\infty \leq e_{\max,\mathrm{D}}$ is satisfied for all times t. Note that for the event-based state feedback with ideal communication the bound

$$e_{\max} = 2.8$$

results (Eq. (3.32)). In order to give the same approximation guarantees in the non-ideal case, the threshold $\bar{e}$ has to be reduced which generally invokes a more frequent communication (cf. Section 3.4.3).

5.2.7. Disturbance estimation

This section extends the previous results by including a disturbance estimate $\hat{\boldsymbol{d}}_k$ of the disturbance $\boldsymbol{d}(t)$ in the control input generator (cf. Eq. (3.24), page 35). In the time interval $[t_k + \tau_k, t_{k+1} + \tau_{k+1})$, the control input $\boldsymbol{u}(t)$ is generated by the model

$$\dot{\boldsymbol{x}}_{\mathrm{s}}(t) = \bar{\boldsymbol{A}}\boldsymbol{x}_{\mathrm{s}}(t) + \boldsymbol{E}\hat{\boldsymbol{d}}_k, \quad \boldsymbol{x}_{\mathrm{s}}(t_k + \tau_k^+) = \boldsymbol{x}_{\mathrm{s}k^+} \tag{5.27}$$

$$\boldsymbol{u}(t) = -\boldsymbol{K}\boldsymbol{x}_{\mathrm{s}}(t). \tag{5.28}$$

However, due to the delay of the communication network, the previous disturbance estimation (3.23), (3.24) has to be adapted since the control input generator gets the information about the plant state $\boldsymbol{x}(t_k)$ with the delay τ_k. There are at least two possibilities:

- Both components work with different disturbance estimates.
- The event generator uses the established algorithm

$$\hat{\boldsymbol{d}}_0 = \boldsymbol{0} \tag{5.29}$$

$$\hat{\boldsymbol{d}}_k = \hat{\boldsymbol{d}}_{k-1} + \left(\boldsymbol{A}^{-1}\left(\mathrm{e}^{\boldsymbol{A}(t_k - t_{k-1})} - \boldsymbol{I}_n\right)\boldsymbol{E}\right)^+ \left(\boldsymbol{x}(t_k) - \boldsymbol{x}_{\mathrm{e}}(t_k^-)\right) \tag{5.30}$$

  to determine the disturbance estimate $\hat{\boldsymbol{d}}_k$. At event time t_k, the event generator communicates this estimate together with the state information $\boldsymbol{x}(t_k)$ and the time stamp t_k towards the control input generator (Fig. 5.5).

The following investigation uses the second possibility.

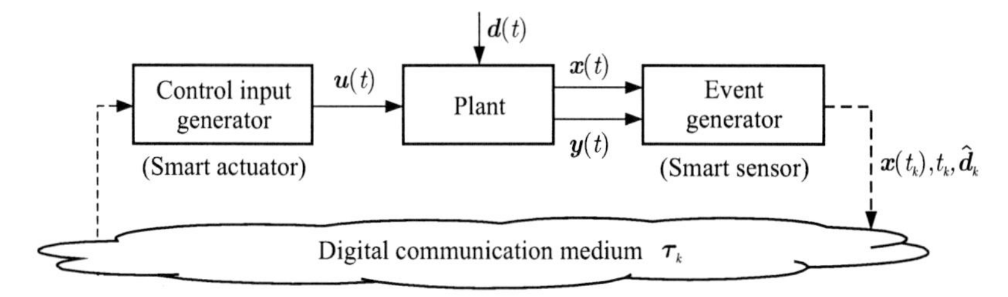

Figure 5.5.: Event-based state-feedback loop with delays and disturbance estimation

Determination of $\boldsymbol{x}_{\mathrm{s}k^+}$. In the time interval $[t_k, t_{k+1})$, the event generator uses the model

$$\dot{\boldsymbol{x}}_{\mathrm{e}}(t) = \bar{\boldsymbol{A}}\boldsymbol{x}_{\mathrm{e}}(t) + \boldsymbol{E}\hat{\boldsymbol{d}}_k, \quad \boldsymbol{x}_{\mathrm{e}}(t_k^+) = \boldsymbol{x}(t_k), \tag{5.31}$$

whose behaviour can be described by

$$\boldsymbol{x}_{\mathrm{e}}(t) = \mathrm{e}^{\bar{\boldsymbol{A}}(t-t_k)}\boldsymbol{x}(t_k) + \bar{\boldsymbol{A}}^{-1}\left(\mathrm{e}^{\bar{\boldsymbol{A}}(t-t_k)} - \boldsymbol{I}_n\right)\boldsymbol{E}\hat{\boldsymbol{d}}_k. \tag{5.32}$$

At time $t_k + \tau_k$, the state $\boldsymbol{x}_{\mathrm{e}}(t)$ of the event generator is given by

$$\boldsymbol{x}_{\mathrm{e}}(t_k+\tau_k) = \mathrm{e}^{\bar{\boldsymbol{A}}\tau_k}\boldsymbol{x}(t_k) + \bar{\boldsymbol{A}}^{-1}\left(\mathrm{e}^{\bar{\boldsymbol{A}}\tau_k} - \boldsymbol{I}_n\right)\boldsymbol{E}\hat{\boldsymbol{d}}_k.$$

The control input generator described by Eqs. (5.27), (5.28) gets the information $\boldsymbol{x}(t_k)$, $\hat{\boldsymbol{d}}_k$ and t_k at the time $t = t_k + \tau_k$. Therefore, it knows the delay $\tau_k = t - t_k$ and a synchronous behaviour of the models of the event generator and the control input generator can be recovered by using

$$\boldsymbol{x}_{\mathrm{s}k^+} = \mathrm{e}^{\bar{\boldsymbol{A}}\tau_k}\boldsymbol{x}(t_k) + \bar{\boldsymbol{A}}^{-1}\left(\mathrm{e}^{\bar{\boldsymbol{A}}\tau_k} - \boldsymbol{I}_n\right)\boldsymbol{E}\hat{\boldsymbol{d}}_k = \boldsymbol{x}_{\mathrm{e}}(t_k+\tau_k). \tag{5.33}$$

In summary, the event-based state-feedback loop consists of

- the plant (5.1), (5.2),
- the control input generator (5.27), (5.28) using the state update mechanism (5.33), and
- the event generator (5.8), (5.31) which estimates the disturbance by means of the recursion (5.29), (5.30).

At event times t_k $(k = 1, 2, ...)$, the measured state $\boldsymbol{x}(t_k)$, the event time t_k and the disturbance estimate $\hat{\boldsymbol{d}}_k$ are sent from the event generator towards the control input generator.

Main result. The analysis of Sections 5.2.4–5.2.6 also holds for the situation considered here. However, as a consequence of the disturbance estimation, the size of the error bound specified in Theorem 15 has to be recalculated.

Lemma 15. *Assume that the transformed disturbance $\boldsymbol{d}_\Delta(t) = \boldsymbol{d}(t) - \hat{\boldsymbol{d}}_{k-1}$ is bounded according to*

$$\|\boldsymbol{d}_\Delta(t)\| = \|\boldsymbol{d}(t) - \hat{\boldsymbol{d}}_{k-1}\| \le \gamma\, d_{\max} \quad \textit{for } t \ge t_{k-1} \tag{5.34}$$

($0 \le \gamma \le 2$, cf. Eq. (3.38), page 47). Moreover, assume that the maximum delay of the communication network is bounded from above by $\tau \le \bar{\tau} \le \tau^{\star\mathrm{d}}$, where $\tau^{\star\mathrm{d}}$ satisfies the condition

$$\tau^{\star\mathrm{d}} = \arg\min_{\tau\in[0,\bar{\tau}]}\left\{\frac{\int_0^\tau\left(\left\|\mathrm{e}^{\boldsymbol{A}\alpha}\right\| + \left\|\mathrm{e}^{\bar{\boldsymbol{A}}\alpha}\right\|\right)\mathrm{d}\alpha\cdot\|\boldsymbol{E}\|\gamma\, d_{\max}}{1 - \left\|\mathrm{e}^{\boldsymbol{A}\tau} - \mathrm{e}^{\bar{\boldsymbol{A}}\tau}\right\|} = \bar{e}\right\} \tag{5.35}$$

with

$$\tilde{\tau} = \arg\min_{\tau \geq 0} \left\{ \left\| e^{\boldsymbol{A}\tau} - e^{\bar{\boldsymbol{A}}\tau} \right\| = 1 \right\}. \tag{5.36}$$

Then, the approximation error $\boldsymbol{e}(t) = \boldsymbol{x}(t) - \boldsymbol{x}_{\mathrm{CT}}(t)$ *between the state* $\boldsymbol{x}(t)$ *of the event-based state-feedback loop* (5.1), (5.2), (5.8), (5.27)–(5.31), (5.33) *and the state* $\boldsymbol{x}_{\mathrm{CT}}(t)$ *of the continuous-time state-feedback loop* (5.4), (5.5) *is bounded from above by*

$$\|\boldsymbol{e}(t)\| \leq e_{\max,\mathrm{Dd}}(\bar{\tau})$$

with

$$e_{\max,\mathrm{Dd}}(\bar{\tau}) = x_{\Delta\max,\mathrm{d}}(\bar{\tau}) \cdot \int_0^\infty \left\| e^{\bar{\boldsymbol{A}}\alpha} \boldsymbol{B}\boldsymbol{K} \right\| \, \mathrm{d}\alpha$$

and

$$x_{\Delta\max,\mathrm{d}}(\bar{\tau}) = \max_{\tau \in [0,\, \bar{\tau}]} \left\| e^{\boldsymbol{A}\tau} \right\| \bar{e} + \int_0^{\bar{\tau}} \left\| e^{\boldsymbol{A}\alpha} \boldsymbol{E} \right\| \, \mathrm{d}\alpha \cdot \gamma\, d_{\max}. \tag{5.37}$$

Proof. See Appendix B.11, page 182. □

The effect of the disturbance estimation is given by the fact that instead of the bound $d_{\max}$ (Eq. (5.19)) the disturbance bound $\gamma\, d_{\max}$ appears in Eqs. (5.35), (5.37). If $\hat{\boldsymbol{d}}_k$ is a good disturbance estimate for the time interval $t \geq t_k$, it holds $\gamma < 1$, and the bound derived here is smaller than the bound of the event-based control loop that works without a disturbance estimation. However, the bound increases for quickly varying disturbances ($\gamma > 1$).

This likewise holds for the admissible delay according to conditions (5.35), (5.36). Note that even for $\gamma = 1$, the bound $\tau^{\star\mathrm{d}}$ is smaller than $\tau^{\star}$ (see conditions (5.22), (5.35)) because the disturbance estimation also reduces the minimum inter-event time for quickly varying disturbances as discussed in Section 3.4.3.

Discussion. The disturbance estimation uses the formula (5.30) for determining the disturbance estimate. It assumes that the disturbance is constant, $\boldsymbol{x}_{\Delta}(t_k)$ is zero and $\boldsymbol{x}_{\mathrm{e}}(t) = \boldsymbol{x}_{\mathrm{s}}(t)$ holds for all times t. However, when considering delays the latter two assumptions are generally violated (Eq. (5.14)).

As a consequence, recursion (5.30) estimates not only the influence of the disturbance but also the influence of the "old" input generated by the control input generator in the time interval

$[t_k, t_k + \tau_k)$. Hence, the disturbance estimation is only slightly affected for small delays but, for large delays, the estimation error increases.

In order to exclude this influence, the disturbance estimation should be only applied for the time interval $t_k + \tau_k \leq t < t_{k+1}$ in which the states of the event generator and the control input generator are identical ($\boldsymbol{x}_{\mathrm{e}}(t) = \boldsymbol{x}_{\mathrm{s}}(t)$). Moreover, $\boldsymbol{x}_{\Delta}(t_k + \tau_k) \neq \boldsymbol{0}$ has to be taken into account:

$$\begin{aligned}
&\hat{\boldsymbol{d}}_0 = \boldsymbol{0}\\
&\hat{\boldsymbol{d}}_k = \hat{\boldsymbol{d}}_{k-1} + \left(\boldsymbol{A}^{-1}\left(\mathrm{e}^{\boldsymbol{A}(t_k - t_{k-1} - \tau_{k-1})} - \boldsymbol{I}_n\right)\boldsymbol{E}\right)^{+} \cdot\\
&\left(\boldsymbol{x}(t_k) - \boldsymbol{x}_{\mathrm{e}}(t_k^-) - \mathrm{e}^{\boldsymbol{A}(t_k - t_{k-1} - \tau_{k-1})}(\boldsymbol{x}(t_{k-1} + \tau_{k-1}) - \boldsymbol{x}_{\mathrm{e}}(t_{k-1} + \tau_{k-1}))\right).
\end{aligned}$$

This recursion requires that the event generator knows the communication delay τ_{k-1} at event time t_k which, however, is only available in the control input generator. In order to overcome this problem, the control input generator has to send, at time $t_{k-1} + \tau_{k-1}$, an acknowledgement signal over a reliable channel to the event generator. This concept is discussed in the next section in the context of packet losses.

5.3. Packet losses

5.3.1. Modification of the event condition

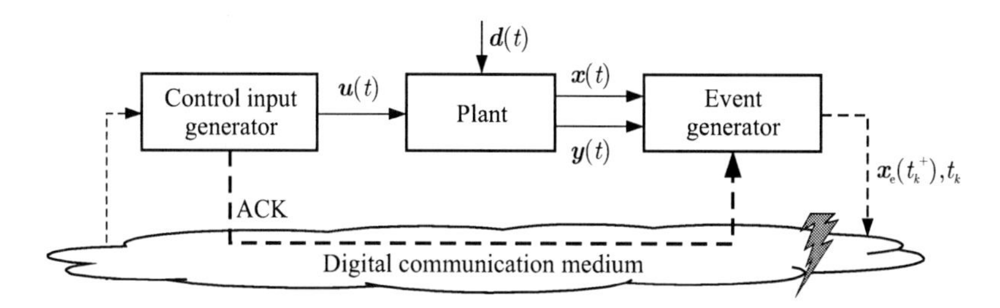

Figure 5.6.: Event-based state-feedback loop subject to packet losses

This section proposes an extended event-triggering mechanism which uses *acknowledgement signals* (ACK) to better cope with packet losses in the event-based control loop. The acknowledgement signals are sent from the control input generator to the event generator over a reliable channel if the information $(\boldsymbol{x}_e(t_k^+), t_k)$ has successfully arrived at the control input generator (Fig. 5.6).

The extended mechanism is illustrated in Fig. 5.7. At event time t_k, at which the event condition $\|\boldsymbol{x}(t_k) - \boldsymbol{x}_e(t_k^-)\| = \bar{e}$ is satisfied, the data sent by the event generator was dropped. Therefore, only the state $\boldsymbol{x}_e(t_k^-)$ of the event generator is updated with the current measurement $\boldsymbol{x}(t_k)$. As the event generator did not get any acknowledgement signal from the control input generator after a predefined waiting time, which is denoted by $T_w \in \mathbb{R}_+$, it resends the information $(\boldsymbol{x}_e(t_k + T_w), t_{k+1})$ at the new event time $t_{k+1} = t_k + T_w$. The state $\boldsymbol{x}_s(t_{k+1}^-)$ of the control input generator is updated with the current state $\boldsymbol{x}_e(t_{k+1})$ of the event generator and both states coincide for $t > t_{k+1}$.

The event-triggering mechanism works as follows:

An event is generated if

$$\begin{cases} t - t_{k-1} = T_w, & \textit{if the transmission has failed at time } t_{k-1} \textit{ (no ACK arrived)} \\ \|\boldsymbol{x}(t) - \boldsymbol{x}_e(t)\| = \bar{e}, & else. \end{cases} \tag{5.38}$$

In both cases, the time t denotes the new event time t_k.

At the event times t_k $(k = 1, 2, ...)$, the model state $\boldsymbol{x}_e(t_k)$ of the event generator is updated

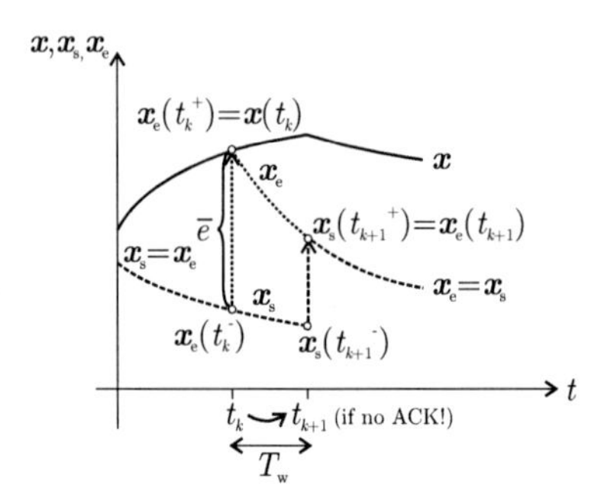

Figure 5.7.: Behaviour of the event-based state-feedback loop subject to packet losses

according to

$$\boldsymbol{x}_{\mathrm{e}}(t_k^+) = \begin{cases} \boldsymbol{x}(t_k), & \text{if } \|\boldsymbol{x}(t_k) - \boldsymbol{x}_{\mathrm{e}}(t_k^-)\| = \bar{e} \\ \boldsymbol{x}_{\mathrm{e}}(t_k), & \text{else} \end{cases} \tag{5.39}$$

and the information $(\boldsymbol{x}_{\mathrm{e}}(t_k^+), t_k)$ is sent towards the control input generator. The state $\boldsymbol{x}_{\mathrm{e}}(t_k^+)$ is used to update the model state $\boldsymbol{x}_{\mathrm{s}}(t_k)$ of the control input generator so that both models have an identical behaviour for all times $t \in [t_k, t_{k+1})$ if the transmission was successful.

Note that for an instantaneous information transfer and a negligible waiting time $T_{\mathrm{w}} \approx 0$, the behaviour of the event-based control loop with the event generating mechanism (5.38) is not affected by data losses because new information is resent immediately until the data packet has been successfully transmitted to the control input generator.

5.3.2. Packet losses and communication delays

Packet losses usually result from a high traffic load of the digital network which likewise causes communication delays. Thus, packet losses and communication delays generally have to be considered simultaneously. Next, the behaviour of the event-based control loop affected by packet losses and communication delays is investigated.

Figure 5.8 shows the scenario in which at the event times t_k and t_{k+1} the information transmission has failed which causes a significant communication delay after event time t_k. Using the extended event generating mechanism (5.38) and the state updating according to Eq. (5.39), the event generator successfully sends the information $(\boldsymbol{x}_{\mathrm{e}}(t_{k+2}^+), t_{k+2})$ at the event time t_{k+2}.

However, due to the communication delay τ_{k+2}, the control input generator gets this information at the time $t_{k+2} + \tau_{k+2}$ at which its state $\boldsymbol{x}_{\mathrm{s}}(t_{k+2} + \tau_{k+2})$ is updated according to the

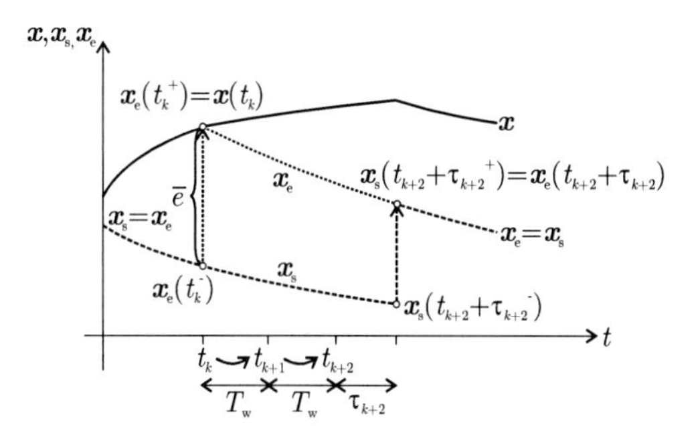

Figure 5.8.: Behaviour of the event-based state-feedback loop subject to packet losses and delays

update rule (5.12):

$$\boldsymbol{x}_{\mathrm{s}k+2^+} = \mathrm{e}^{\bar{\boldsymbol{A}}\tau_{k+2}} \boldsymbol{x}_{\mathrm{e}}(t_{k+2}^+).$$

In this example, the overall communication delay $\tau_{\mathrm{o}k}$, which occurs after event time t_k, is given by

$$\tau_{\mathrm{o}k} = T_{\mathrm{w}} + T_{\mathrm{w}} + \tau_{k+2}.$$

Note that in order not to send information unnecessarily, the waiting time T_{w} must be larger than the maximum possible communication delay $T_{\mathrm{w}} > \bar{\tau}$ but also smaller than the minimum time interval between two consecutive events $T_{\mathrm{w}} < t_{k+1} - t_k, \;\; \forall\, k$ (Section 5.2.5).

Under the assumption that the maximum number of consecutive packet losses N_{consec} is bounded by $N_{\mathrm{consec}} \leq \bar{N}$ and the transmission of the acknowledgement signal is instantaneous, T_{w} can be set to $\bar{\tau}$ and the analysis of Sections 5.2.4–5.2.6 can be simply adopted to get the following result.

Corollary 2. *Consider the event-based control loop* (5.1), (5.2), (5.7), (5.10)–(5.12), (5.38), (5.39) *affected by communication delays and packet losses. Assume that the maximum communication delay of the communication network is bounded from above by*

$$\bar{\tau} \leq \frac{\tau^\star}{\bar{N}+1},$$

where $\tau^\star$ results from the conditions (5.22), (5.23) *and $\bar{N}$ is an upper bound for the maximum number of consecutive packet losses. Then, the difference $\mathbf{e}(t) = \boldsymbol{x}(t) - \boldsymbol{x}_{\mathrm{CT}}(t)$ between the state $\boldsymbol{x}(t)$ of the event-based state-feedback loop and the state $\boldsymbol{x}_{\mathrm{CT}}(t)$ of the continuous-time state-feedback loop* (5.4), (5.5) *without communication imperfections is bounded from above* *by*

$$\|\boldsymbol{e}(t)\| \leq e_{\max,\mathrm{D}}(\bar{\tau})$$

with

$$e_{\max,\mathrm{D}}(\bar{\tau}) = x_{\Delta\max}(\bar{\tau}) \cdot \int_0^\infty \left\| \mathrm{e}^{\bar{\boldsymbol{A}}\alpha} \boldsymbol{B}\boldsymbol{K} \right\| \, \mathrm{d}\alpha$$

and $x_{\Delta\max}(\bar{\tau})$ *according to Eq.* (5.19).

The only difference to the result stated in Theorem 15 is that the assumption $\bar{\tau} \leq \tau^\star$ is replaced by $\bar{\tau} \leq \frac{\tau^\star}{\bar{N}+1}$. Hence, the admissible delay $\bar{\tau}$ of the communication network decreases for $\bar{N} > 0$. For $\bar{N} = 0$, which describes the communication network without packet losses, Corollary 2 and Theorem 15 are identical.

5.4. Data-rate constraints

5.4.1. Problem description

Besides communication delays and packet losses, the digital communication network imposes a limited bandwidth, which may restrict the information content to be sent to a small number of bits.

To allow a quantised communication at event times t_k [4, 23], this section extends the event-based state feedback by incorporating the quantisation scheme proposed in [76]. The crucial difference to the approach studied in [76], which deals with discrete-time systems, is that the quantisation scheme in now adapted to the needs of event-based control. The main results of this section are the following:

- By including a suitable encoder and decoder in the event-based control loop, the state $\boldsymbol{x}(t)$ of the quantised event-based state-feedback loop remains *GUUB* and an approximation error bound can be calculated (Theorem 16).
- There exists a minimum inter-event time which explicitly depends on the quantisation (Theorems 17, 18).

5.4.2. Structure of the quantised event-based state-feedback loop

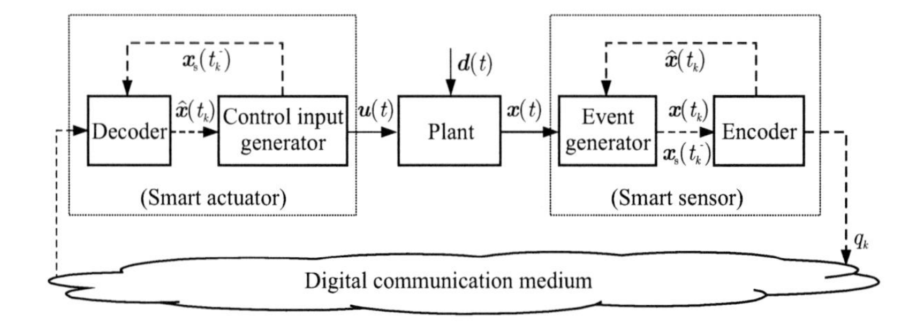

Figure 5.9.: Quantised event-based state-feedback loop

The quantised event-based state-feedback loop is depicted in Fig. 5.9. In comparison to the original loop (Fig. 5.1), the smart actuator and the smart sensor include not only the control input generator and the event generator but also an encoder and a decoder. The encoder has the task to quantise the state information $\boldsymbol{x}(t_k)$ obtained from the event generator at event time

t_k, whereas the decoder has the task to translate the quantised information $q_k \in \mathbb{N}$ obtained at time t_k from the encoder into approximate state information $\hat{\boldsymbol{x}}(t_k)$.

In this section, the communication network is assumed to not induce any delays or packet losses as long as its specific data rate $R \in \mathbb{N}_{>0}$ (given in bits/sample) is matched by the information (user data) sent at event time.

Control input generator and event generator. In the time interval $[t_k, t_{k+1})$ between two consecutive events, both the control input generator and the event generator run the model

$$\dot{\boldsymbol{x}}_{\mathrm{s}}(t) = \bar{\boldsymbol{A}}\boldsymbol{x}_{\mathrm{s}}(t), \ \boldsymbol{x}_{\mathrm{s}}(t_k^+) = \hat{\boldsymbol{x}}(t_k) \tag{5.40}$$

$$\boldsymbol{u}(t) = -\boldsymbol{K}\boldsymbol{x}_{\mathrm{s}}(t) \tag{5.41}$$

to produce the control input $\boldsymbol{u}(t)$ and to determine the event times t_k. Note that both models are synchronously updated at these event times with the same information $\hat{\boldsymbol{x}}(t_k)$ and, hence, both models have an identical behaviour for all times $t \geq 0$.

An event is generated whenever the difference between the measured state $\boldsymbol{x}(t)$ and the model state $\boldsymbol{x}_{\mathrm{s}}(t)$ reaches the event threshold $\bar{e}$

$$\|\boldsymbol{x}(t) - \boldsymbol{x}_{\mathrm{s}}(t)\|_\infty = \bar{e}, \quad t := t_k. \tag{5.42}$$

At event time t_k, the quantised information q_k generated by the encoder is sent from the smart sensor towards the smart actuator and is used there as well as in the sensor to get the approximate state $\hat{\boldsymbol{x}}(t_k)$.

Throughout this section, the supremum norm $\| \star \|_\infty$ is used both for the event generation and the analysis due to the quantisation method applied. However, an adaptation to arbitrary norms is possible but increases the analytical complexity.

In the following, first, the problems of deriving a suitable quantised information q_k and determining a suitable approximate state $\hat{\boldsymbol{x}}(t_k)$ are addressed. Later, Section 5.4.7 discusses the problems which arise with respect to incorporating a disturbance estimation.

5.4.3. Description of the encoder and decoder

In order to avoid sending the full state information $\boldsymbol{x}(t_k)$ at event times and to restrict the information sent to a limited number of bits, the encoder is included into the feedback loop for quantising the state information $\boldsymbol{x}(t_k)$

$$\mathrm{enc} : \mathbb{R}^n \to \mathbb{N}_{>0}.$$

The decoder, which obtains the quantised state information $q_k \in \mathbb{N}_{>0}$ from the encoder at event times t_k, is used to decode this information

$$\mathrm{dec} : \mathbb{N}_{>0} \to \mathbb{R}^n.$$

Algorithm 1. *Assume that initially* $\|\boldsymbol{x}(0) - \boldsymbol{x}_{\mathrm{s}}(0)\|_\infty < \bar{e}$ *holds. At each event time* t_k $(k = 1, 2, ...)$, *the quantisation algorithm works as follows.*

Task of the encoder:

1. *At event time* t_k, *the encoder processes the plant state* $\boldsymbol{x}(t_k)$ *and the model state* $\boldsymbol{x}_{\mathrm{s}}(t_k^-)$. *It uses the model state* $\boldsymbol{x}_{\mathrm{s}}(t_k^-)$ *as the centre point of the region*

$$\Omega(\boldsymbol{x}_{\mathrm{s}}(t_k^-)) = \{\boldsymbol{x} : \|\boldsymbol{x} - \boldsymbol{x}_{\mathrm{s}}(t_k^-)\|_\infty \leq \bar{e}\}.$$

Due to event condition (5.42), *the state* $\boldsymbol{x}(t_k)$ *always lies on the boundary*

$$\partial\Omega(\boldsymbol{x}_{\mathrm{s}}(t_k^-)) = \{\boldsymbol{x} : \|\boldsymbol{x} - \boldsymbol{x}_{\mathrm{s}}(t_k^-)\|_\infty = \bar{e}\}$$

of this set.

2. *The region* $\Omega(\boldsymbol{x}_{\mathrm{s}}(t_k^-))$ *is subdivided by the encoder into* N *equal subregions* $\mathcal{P}_k^q \subset \mathbb{R}^n$ *numbered with* $q \in \{0, 1, ..., N-1\}$, *where* $N = \xi^n$, *and* $\xi \in \{2, 3, ...\}$ *denotes the quantisation level.*

3. *The number* q_k *of the subregion* $\mathcal{P}_k^q$, *which contains the plant state at event time* t_k, *i.e.* $\boldsymbol{x}(t_k) \in \mathcal{P}_k^q$, *and the centre point* $\hat{\boldsymbol{x}}(t_k)$ *of this subregion are determined by the encoder. Finally,* q_k *is sent over the network to the decoder and* $\hat{\boldsymbol{x}}(t_k)$ *is used to update the model in the event generator.*

Task of the decoder:

4. *By means of* q_k, *the predefined event threshold* $\bar{e}$, *the fixed maximum number* N *of subregions, and the state* $\boldsymbol{x}_{\mathrm{s}}(t_k^-)$ *provided by the control input generator (Fig. 5.9), the decoder also determines the centre point* $\hat{\boldsymbol{x}}(t_k)$ *of the subregion* $\mathcal{P}_k^q$. *The resulting quantisation error* $\|\boldsymbol{x}(t_k) - \hat{\boldsymbol{x}}(t_k)\|_\infty$ *is given by*

$$\|\boldsymbol{x}(t_k) - \hat{\boldsymbol{x}}(t_k)\|_\infty = \frac{\bar{e}}{\xi}. \tag{5.43}$$

5. *The decoder provides the control input generator with the state $\hat{\boldsymbol{x}}(t_k)$ which is used there to update the model state $\boldsymbol{x}_{\mathrm{s}}(t_k)$*

$$\boldsymbol{x}_{\mathrm{s}}(t_k^+) = \hat{\boldsymbol{x}}(t_k) \tag{5.44}$$

(Fig. 5.9 and Eq. (5.40)*).*

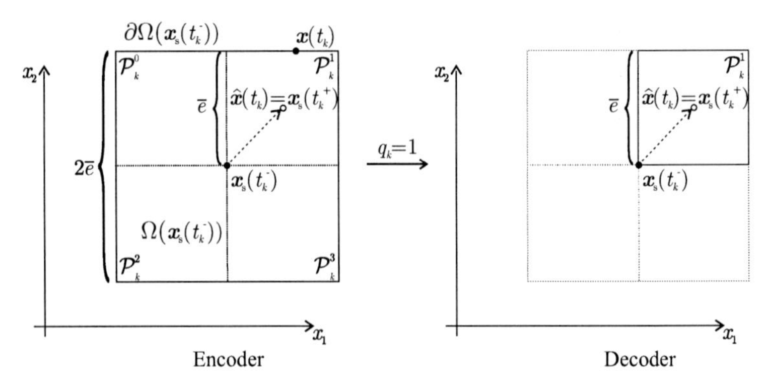

Figure 5.10.: Quantisation mechanism

Figure 5.10 illustrates the quantisation mechanism for a second-order system and $N = 4$. At the event time t_k, the encoder (left-hand side of the figure) determines in which partition the measured state $\boldsymbol{x}(t_k)$ is located, i.e. $\boldsymbol{x}(t_k) \in \mathcal{P}_k^1$, and sends the information $q_k = 1$ to the decoder (right-hand side of the figure). The decoder gets the state $\boldsymbol{x}_{\mathrm{s}}(t_k^-)$ from the control input generator and determines the approximate state $\hat{\boldsymbol{x}}(t_k)$ according to the information $q_k = 1$ and the knowledge about $\bar{e}$ and N. The state $\hat{\boldsymbol{x}}(t_k)$ is also determined by the encoder and used in the event generator and in the control input generator to update their model states according to Eq. (5.44).

In the event-based scenario considered, the maximum number of partitions N can be reduced to a maximum number of relevant partitions N_{R} because, at event times t_k,

$$\boldsymbol{x}(t_k) \in \partial\Omega(\boldsymbol{x}_{\mathrm{s}}(t_k^-))$$

holds (Eq. (5.42)). Therefore, only the partitions which are adjacent to the boundary $\partial\Omega(\boldsymbol{x}_{\mathrm{s}}(t_k^-))$ have to be taken into account, the number of which is given by

$$N_{\mathrm{R}} = N - (\xi - 2)^n = \xi^n - (\xi - 2)^n \tag{5.45}$$

(left-hand side of Fig. 5.11). The number N_{R} is associated with the maximum number of bits which may be sent over the network in a single packet according to the data rate R of the network

$$R = \lceil \log_2 N_{\mathrm{R}} \rceil$$

[93], where $\lceil \star \rceil$ denotes the upper integer of $\star$.

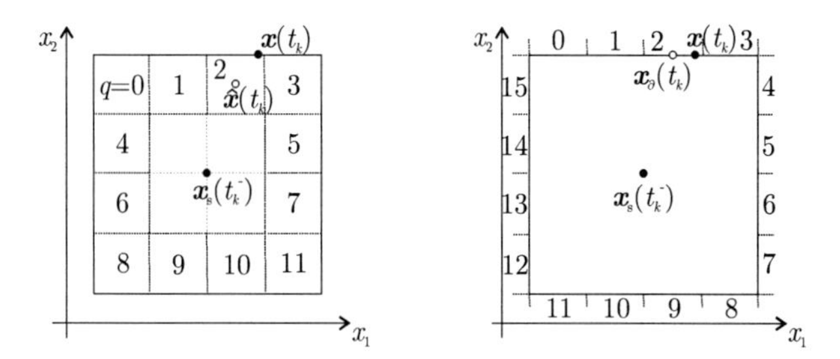

Figure 5.11.: Relevant partitions for $\xi = 4$ and $n = 2$

Instead of using the approximate state $\hat{\boldsymbol{x}}(t_k)$ for updating the model states $\boldsymbol{x}_{\mathrm{s}}(t)$, the state $\boldsymbol{x}_\partial(t_k)$ can be used which describes the centre point of the corresponding partition on the boundary $\partial\Omega(\boldsymbol{x}_{\mathrm{s}}(t_k^-))$ (right-hand side of Fig. 5.11). The advantage of this quantisation is that for the same quantisation level ξ, the quantisation error is generally smaller since one state variable is exactly determined. Hence, inequality

$$\|\boldsymbol{x}(t_k) - \boldsymbol{x}_\partial(t_k)\|_\infty \leq \frac{\bar{e}}{\xi}$$

holds instead of Eq. (5.43). The disadvantage of this mechanism lies in an increased number of bits to be sent at event times since the number of partition elements

$$N_{\partial\mathrm{R}} = N_{\mathrm{R}} + 2^n(n-1)$$

is larger due to the splitting of the partitions at the corners, which is also depicted in Fig. 5.11. As the additive term is independent of the quantisation level ξ, it becomes negligible for large ξ. However, for large ξ also the benefit with respect to the smaller approximation error decreases. Note that the splitting problem can be avoided by using e.g the Euclidean norm.

The following analysis concentrates on the update with the approximate state $\hat{\boldsymbol{x}}(t_k)$.

Summary of the components. The quantised event-based state-feedback loop consists of

- the plant (5.1), (5.2),
- the smart actuator incorporating the event generator (5.40), (5.42) and the decoder (Algorithm 1), and
- the smart sensor incorporating the control input generator (5.40), (5.41) and the encoder (Algorithm 1).

At event times t_k $(k = 1, 2, ...)$, the quantised state information q_k is sent from the encoder to the decoder.

To guarantee a synchronous behaviour of the models used in the control input generator and the event generator, both models are updated with the approximate state information $\hat{\boldsymbol{x}}(t_k)$ (Fig. 5.9), which restricts the uncertainty about the actual plant state $\boldsymbol{x}(t_k)$ according to relation (5.43). However, due to the quantisation, the control input generator generally does not know the exact state information $\boldsymbol{x}(t)$ for determining the control input $\boldsymbol{u}(t)$. The consequences for the behaviour of the event-based state-feedback loop are investigated in the following.

> The analysis will show that the performance bound on the event-based state-feedback loop is retained if instead of the precise information $\boldsymbol{x}(t_k)$ only the quantised information q_k is communicated at event times t_k (Theorem 16). However, the prize for sending less information lies in smaller bounds for the minimum inter-event time (Theorems 17, 18).

Proposition 7. *For $\xi \in \mathbb{N}_{\geq 2}$, the event generating mechanism* (5.42) *and Algorithm 1 ensure that the state difference $\boldsymbol{x}_\Delta(t) = \boldsymbol{x}(t) - \boldsymbol{x}_s(t)$ is bounded according to*

$$\|\boldsymbol{x}_\Delta(t)\|_\infty = \|\boldsymbol{x}(t) - \boldsymbol{x}_s(t)\|_\infty \leq \bar{e}, \quad \forall t \geq 0.$$

Proof. Whenever the state $\boldsymbol{x}(t)$ touches the boundary $\partial\Omega(\boldsymbol{x}_s(t))$ of the set $\Omega(\boldsymbol{x}_s(t))$, i.e. $\|\boldsymbol{x}(t) - \boldsymbol{x}_s(t)\|_\infty = \bar{e}$, the communication mechanism resets the states $\boldsymbol{x}_s(t_k)$ in the event generator and the control input generator according to $\boldsymbol{x}_s(t_k^+) = \hat{\boldsymbol{x}}(t_k)$. Since

$$\|\boldsymbol{x}(t_k) - \hat{\boldsymbol{x}}(t_k)\|_\infty = \frac{\bar{e}}{\xi} < \bar{e}$$

holds for $\xi \in \mathbb{N}_{\geq 2}$ according to Eq. (5.43), $\|\boldsymbol{x}(t) - \boldsymbol{x}_s(t)\|_\infty \leq \bar{e}$ holds for all times t. □

5.4.4. Behaviour of the quantised event-based control loop

The plant (5.1), (5.2) together with the control input generator (5.40), (5.41) is described for the time period $[t_k, t_{k+1})$ by the state-space model

$$\begin{pmatrix} \dot{\boldsymbol{x}}(t) \\ \dot{\boldsymbol{x}}_{\rm s}(t) \end{pmatrix} = \begin{pmatrix} \boldsymbol{A} & -\boldsymbol{BK} \\ \boldsymbol{O} & \bar{\boldsymbol{A}} \end{pmatrix} \begin{pmatrix} \boldsymbol{x}(t) \\ \boldsymbol{x}_{\rm s}(t) \end{pmatrix} + \begin{pmatrix} \boldsymbol{E} \\ \boldsymbol{O} \end{pmatrix} \boldsymbol{d}(t)$$

$$\begin{pmatrix} \boldsymbol{x}(t_k) \\ \boldsymbol{x}_{\rm s}(t_k^+) \end{pmatrix} = \begin{pmatrix} \boldsymbol{x}(t_k) \\ \hat{\boldsymbol{x}}(t_k) \end{pmatrix}.$$

This model can be transformed into

$$\begin{pmatrix} \dot{\boldsymbol{x}}_\Delta(t) \\ \dot{\boldsymbol{x}}_{\rm s}(t) \end{pmatrix} = \begin{pmatrix} \boldsymbol{A} & \boldsymbol{O} \\ \boldsymbol{O} & \bar{\boldsymbol{A}} \end{pmatrix} \begin{pmatrix} \boldsymbol{x}_\Delta(t) \\ \boldsymbol{x}_{\rm s}(t) \end{pmatrix} + \begin{pmatrix} \boldsymbol{E} \\ \boldsymbol{O} \end{pmatrix} \boldsymbol{d}(t)$$

$$\begin{pmatrix} \boldsymbol{x}_\Delta(t_k^+) \\ \boldsymbol{x}_{\rm s}(t_k^+) \end{pmatrix} = \begin{pmatrix} \boldsymbol{x}(t_k) - \hat{\boldsymbol{x}}(t_k) \\ \hat{\boldsymbol{x}}(t_k) \end{pmatrix}$$

according to state transformation (3.14), where $\boldsymbol{x}_\Delta(t_k^+)$ is unknown due to the quantisation but bounded according to relation (5.43). The state $\boldsymbol{x}(t)$ is given by $\boldsymbol{x}(t) = \boldsymbol{x}_{\rm s}(t) + \boldsymbol{x}_\Delta(t)$ with

$$\begin{aligned} \boldsymbol{x}_{\rm s}(t) &= {\rm e}^{\bar{\boldsymbol{A}}(t-t_k)} \hat{\boldsymbol{x}}(t_k) \\ \boldsymbol{x}_\Delta(t) &= {\rm e}^{\boldsymbol{A}(t-t_k)} \boldsymbol{x}_\Delta(t_k^+) + \int_{t_k}^{t} {\rm e}^{\boldsymbol{A}(t-\alpha)} \boldsymbol{E}\boldsymbol{d}(\alpha)\, {\rm d}\alpha. \end{aligned} \tag{5.46}$$

5.4.5. Comparison of the quantised event-based state-feedback loop and the continuous-time state-feedback loop

For the difference $\boldsymbol{e}(t) = \boldsymbol{x}(t) - \boldsymbol{x}_{\rm CT}(t)$ between the states of the quantised event-based state-feedback loop (5.1), (5.2), (5.40)–(5.43) and the continuous-time state-feedback loop (5.4), (5.5), the relation

$$\dot{\boldsymbol{e}}(t) = \bar{\boldsymbol{A}}\boldsymbol{e}(t) + \boldsymbol{BK}\boldsymbol{x}_\Delta(t), \qquad \boldsymbol{e}(0) = \boldsymbol{0}$$

holds (cf. Section 3.4.1). As the difference state $\boldsymbol{x}_\Delta(t)$ is bounded according to Proposition 7 and the matrix $\bar{\boldsymbol{A}}$ is assumed to be Hurwitz, the following result is obtained.

Theorem 16. *The approximation error* $\mathbf{e}(t) = \boldsymbol{x}(t) - \boldsymbol{x}_{\mathrm{CT}}(t)$ *between the state* $\boldsymbol{x}(t)$ *of the quantised event-based state-feedback loop* (5.1), (5.2), (5.40)–(5.43) *and the state* $\boldsymbol{x}_{\mathrm{CT}}(t)$ *of the continuous-time state-feedback loop* (5.4), (5.5) *is bounded from above by*

$$\|\boldsymbol{e}(t)\|_\infty \leq e_{\max}$$

with

$$e_{\max} = \bar{e} \cdot \int_0^\infty \left\| \mathrm{e}^{\bar{\boldsymbol{A}}\alpha} \boldsymbol{B}\boldsymbol{K} \right\|_\infty \mathrm{d}\alpha. \tag{5.47}$$

The bound $e_{\max}$ derived in this section is the same bound as derived in Theorem 3, where the full state information $\boldsymbol{x}(t_k)$ is sent at event times. It is interesting to note that sending only quantised information does not affect the guaranteed performance but, of course, the actual behaviour of the event-based control loop due to the imprecise information at event times.

Moreover, the theorem shows that the event-based controller with quantised state information is able to arbitrarily approximate the continuous-time state feedback with precise state information by accordingly choosing the event threshold $\bar{e}$. This holds because the event threshold $\bar{e}$ also affects the quantisation according to relation (5.43). For $\bar{e} = 0$, the plant state $\boldsymbol{x}(t)$ can be exactly reconstructed but the communication is invoked continuously.

5.4.6. Minimum inter-event time

In general, a minimum inter-event time $T_{\min} > 0$ exists, because the inequality

$$\|\boldsymbol{x}_\Delta(t_k^+)\|_\infty < \bar{e}$$

holds according to Proposition 7. This section develops a lower bound for the minimum inter-event time $T_{\min}$ both in the case of an undisturbed plant and a plant subject to disturbances.

Undisturbed plant $(\boldsymbol{d}(t) = 0)$**.** As an event is generated if Eq. (5.42) holds, Eq. (5.46) with $\boldsymbol{d}(t) = \mathbf{0}$ leads to the relations

$$\begin{aligned}
\|\boldsymbol{x}_\Delta(t)\|_\infty &= \left\| \mathrm{e}^{\boldsymbol{A}(t-t_k)} \boldsymbol{x}_\Delta(t_k^+) \right\|_\infty \\
&\leq \left\| \mathrm{e}^{\boldsymbol{A}t} \right\|_\infty \cdot \|\boldsymbol{x}(t_k) - \hat{\boldsymbol{x}}(t_k)\|_\infty \\
&= \left\| \mathrm{e}^{\boldsymbol{A}t} \right\|_\infty \cdot \frac{\bar{e}}{\xi}
\end{aligned}$$

which yield, for $\bar{e} \neq 0$, the lower bound $\bar{T} \leq T_{\min}$ according to

$$\begin{aligned} \left\| \mathrm{e}^{\boldsymbol{A}t} \right\|_\infty \cdot \frac{\bar{e}}{\xi} &= \bar{e} \\ \Rightarrow \quad \bar{T} &= \arg\min_t \left\{ \left\| \mathrm{e}^{\boldsymbol{A}t} \right\|_\infty = \xi \right\}. \end{aligned} \tag{5.48}$$

Theorem 17. *For* $\boldsymbol{d}(t) = \boldsymbol{0}$ *and* $\bar{e} \neq 0$*, the minimum inter-event time* $T_{\min}$ *of the quantised event-based state-feedback loop* (5.1), (5.2), (5.40)–(5.43) *is bounded from below by* $\bar{T}$ *given by relation* (5.48).

The quantisation error is considered here as the only reason to communicate. In this situation the event-based control loop has the following properties. First, the minimum inter-event time $T_{\min}$ is independent of the event threshold $\bar{e}$ and depends only on the quantisation level ξ and the system dynamics (Theorem 17). Second, the performance bound $e_{\max}$ is independent of the quantisation level ξ (Theorem 16). Thus, the approximation accuracy and the communication are entirely decoupled and, for a tolerable approximation error of the event-based control loop, which can be ensured by accordingly choosing $\bar{e}$, a desired bound for the communication can be independently adjusted by accordingly choosing ξ.

As the data rate depends on ξ according to $R = \lceil \log_2 N_{\mathrm{R}} \rceil = \lceil \log_2(\xi^n - (\xi - 2)^n) \rceil$, relation (5.48) can be used to find a tradeoff between the maximum communication which may be invoked and the maximum number of bits allowed to be sent over the network at event times.

Disturbed plant $(\boldsymbol{d}(t) \neq \boldsymbol{0})$. In the presence of disturbances, the difference state $\boldsymbol{x}_\Delta(t)$ is given for the time period $[t_k, t_{k+1})$ by

$$\boldsymbol{x}_\Delta(t) = \mathrm{e}^{\boldsymbol{A}(t - t_k)} \boldsymbol{x}_\Delta(t_k^+) + \int_{t_k}^{t} \mathrm{e}^{\boldsymbol{A}(t - \alpha)} \boldsymbol{E}\boldsymbol{d}(\alpha) \,\mathrm{d}\alpha$$

(Eq. (5.46)). A lower bound $\bar{T}_{\mathrm{d}} \leq T_{\min}$ for the minimum inter-event time can be derived similarly to the undisturbed case since the disturbance $\boldsymbol{d}(t)$ is assumed to be bounded according to Eq. (5.3). It holds

$$\begin{aligned} \|\boldsymbol{x}_\Delta(t)\|_\infty &= \left\| \mathrm{e}^{\boldsymbol{A}(t-t_k)} \boldsymbol{x}_\Delta(t_k^+) + \int_{t_k}^{t} \mathrm{e}^{\boldsymbol{A}(t-\alpha)} \boldsymbol{E}\boldsymbol{d}(\alpha)\, \mathrm{d}\alpha \right\|_\infty \\ &\leq \max_{t\geq 0} \left\| \mathrm{e}^{\boldsymbol{A}t} \right\|_\infty \cdot \frac{\bar{e}}{\xi} + \int_0^t \left\| \mathrm{e}^{\boldsymbol{A}\alpha} \boldsymbol{E} \right\|_\infty \mathrm{d}\alpha \cdot d_{\max}. \end{aligned}$$

Accordingly, the lower bound $\bar{T}_{\mathrm{d}} \leq T_{\min}$ is the upper integral bound appearing in the relation

$$\max_{t\geq 0} \left\| \mathrm{e}^{\boldsymbol{A}t} \right\|_\infty \cdot \frac{\bar{e}}{\xi} + \int_0^t \left\| \mathrm{e}^{\boldsymbol{A}\alpha} \boldsymbol{E} \right\|_\infty \mathrm{d}\alpha \cdot d_{\max} = \bar{e}$$

$$\Rightarrow \quad \bar{T}_{\mathrm{d}} = \arg\min_t \left\{ \int_0^t \left\| \mathrm{e}^{\boldsymbol{A}\alpha} \boldsymbol{E} \right\|_\infty \mathrm{d}\alpha = \frac{\bar{e}}{d_{\max}} \cdot \left(1 - \frac{\max_{t\geq 0} \left\| \mathrm{e}^{\boldsymbol{A}t} \right\|_\infty}{\xi} \right) \right\}. \quad (5.49)$$

Theorem 18. *If* $\max_{t\geq 0} \left\| \mathrm{e}^{\boldsymbol{A}t} \right\|_\infty < \xi$ *holds, the minimum inter-event time* $T_{\min}$ *of the quantised event-based state-feedback loop* (5.1), (5.2), (5.40)–(5.43) *is bounded from below by* $\bar{T}_{\mathrm{d}}$ *given by relation* (5.49).

Contrary to Theorem 17, the bound $\bar{T}_{\mathrm{d}}$ depends not only on the quantisation level ξ but also on the event threshold $\bar{e}$ due to the disturbance. For $\xi \to \infty$, the bound $\bar{T}_{\mathrm{d}}$ is equal to the bound derived in Theorem 5 for an unrestricted communication. The condition $\max_{t\geq 0} \left\| \mathrm{e}^{\boldsymbol{A}t} \right\|_\infty < \xi$ has to be satisfied since otherwise the uncertainty about the plant state may immediately cause an event even in the undisturbed case ($\boldsymbol{d}(t) = 0$).

5.4.7. Disturbance estimation

A remaining question concerns the suitable incorporation of a disturbance estimation in the quantised event-based state-feedback loop, which causes the following problems. On the one hand, the disturbance estimator could be incorporated in the event generator. This allows the disturbance estimator to determine constant disturbances exactly because the event generator knows the measured state $\boldsymbol{x}(t)$. However, due to the data-rate constraints of the network, only a quantised disturbance information could be sent at event times. This additionally decreases the communication resources available for the transmission of the quantised state information.

On the other hand, an incorporation of the disturbance estimator in the control input generator would prevent the control loop from sending additional (imprecise) disturbance information. However, in this case the problem lies in the imprecise information $\hat{\boldsymbol{x}}(t_k)$ which the control input generator has for the disturbance estimation at event times t_k. The consequences of the second approach on the disturbance estimation have been investigated in [23].

Example 11 *Quantised event-based control of an unstable plant*

The following example is borrowed from [4]. Consider the unstable plant model

$$\dot{\boldsymbol{x}}(t) = \begin{pmatrix} 0.001 & 0 \\ 0 & -0.002 \end{pmatrix} \boldsymbol{x}(t) + \begin{pmatrix} 0.2 & 0 \\ -0.1 & 0.02 \end{pmatrix} \boldsymbol{u}(t) \tag{5.50}$$

with the initial state $\boldsymbol{x}_0 = (1.5 \quad -1.2)'$*, the state-feedback controller*

$$\boldsymbol{K} = \begin{pmatrix} 0.1 & -0.02 \\ 0.23 & 0.88 \end{pmatrix},$$

quantisation level $\xi = 5$ *and event threshold* $\bar{e} = 2$.

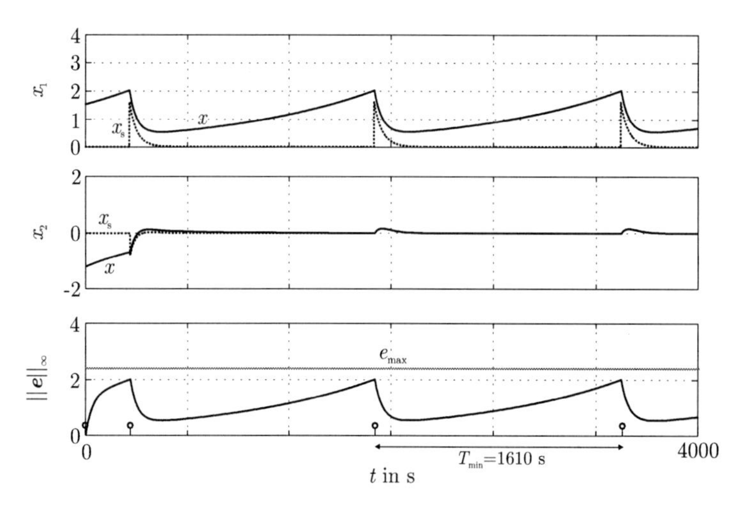

Figure 5.12.: *Behaviour of the quantised event-based control loop with an unstable plant. Solid lines: plant state* $\boldsymbol{x}(t)$*; dotted lines: model state* $\boldsymbol{x}_s(t)$.

The behaviour of the quantised event-based control loop is shown in Fig. 5.12. Due to the initial uncertainty $\boldsymbol{x}_0 \neq \boldsymbol{x}_s(0) = (0 \ \ 0)'$ *and the instability of plant (5.50), the deviation* *of $x_1(t)$ and $x_{s,1}(t)$ increases until the first event at time t_1 which is caused by*

$$|x_1(t_1) - x_{s,1}(t_1)| = 2.$$

At time t_1, the state $\boldsymbol{x}_s(t_1^-)$ is updated with the approximate state $\hat{\boldsymbol{x}}(t_1)$ generated by the decoder. Hence,

$$\|\boldsymbol{x}(t_1) - \boldsymbol{x}_s(t_1^+)\|_\infty = 0.4 \neq 0 \tag{5.51}$$

holds according to Eq. (5.43) so that a new event occurs, which also holds at each further event time t_k $(k > 2)$. The simulation shows that neither the maximum approximation error $e_{\max}$ nor the lower bound for the minimum inter-event time $\bar{T}$ is violated. The bounds $e_{\max}$ and $\bar{T}$ are determined by means of Eqs. (5.47), (5.48), i.e.

$$\begin{aligned} e_{\max} &= 2.43 \\ \bar{T}_{\mathrm{d}} &= 1609 \text{ s}. \end{aligned}$$

Note that the inter-event time at the beginning $(T = t_1 - t_0)$ is smaller than $\bar{T}$. The reason is the initial uncertainty $\|\boldsymbol{x}(0) - \boldsymbol{x}_s(0)\|_\infty = 1.5$ which is greater than the guaranteed uncertainty at the subsequent event times (Eq. (5.51)).

Example 12 *Quantised event-based control of the thermofluid process*

Consider the thermofluid process (2.7), (2.8) with the initial state $\boldsymbol{x}_0 = \boldsymbol{x}_s(0) = (0\ \ 0)'$ subject to the constant disturbance $d(t) = 0.15 = d_{\max}$. The controller is given by Eq. (3.27), the event threshold is chosen to be $\bar{e} = 2$ and the quantisation level is $\xi = 4$.

Equations (5.47), (5.49) yield the bounds

$$\begin{aligned} e_{\max} &= 2.26 \\ \bar{T} &= 70 \text{ s} \end{aligned}$$

(cf. Examples 2, 4, pages 43, 49).

Figure 5.13 shows the resulting trajectories of the quantised event-based state-feedback loop. After the initialising event at time $t = 0$, overall 13 events occur in the time interval considered caused by

$$|x_1(t_k) - x_{s,1}(t_k)| = 2, \quad k = 1, 2, ..., 13.$$

The four particular changes in the behaviour of $x_1(t)$ and $x_2(t)$ at event times t_1, t_5, t_8 and t_{11} result from $\boldsymbol{x}(t_k)$ which lies in different partitions, i.e. $q_{1;5;8;11} = 4$ instead of $q_k = 6$ at the other event times. As expected, the actual approximation error and the minimum inter-event time are lower than the specified bounds (see lower plot of the figure).

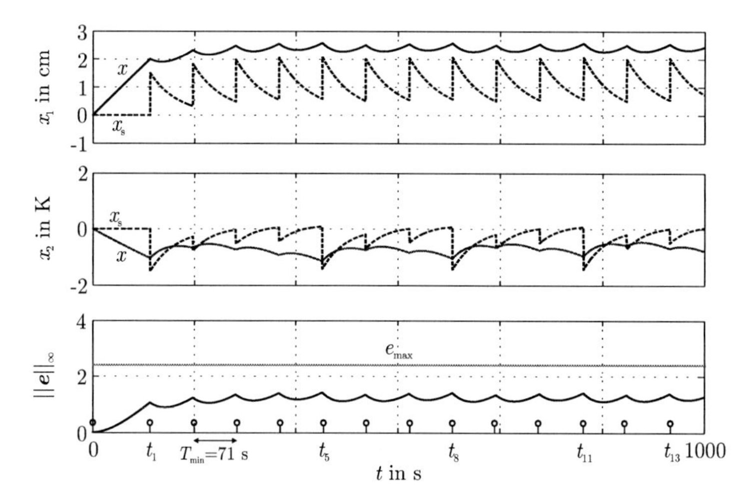

Figure 5.13.: *Behaviour of the quantised event-based control of the thermofluid process. Solid lines: plant state $\boldsymbol{x}(t)$; dotted lines: model state $\boldsymbol{x}_s(t)$.*

6. Experimental evaluation

The experimental evaluation of the event-based state feedback is presented in this chapter. Section 6.1 introduces the chemical pilot plant VERA which has been used to realise the thermofluid process. Section 6.2 discusses implementation aspects related to the event-based state feedback and Section 6.3 shows the experimental results with and without ideal communication properties.

6.1. Chemical pilot plant VERA

Figure 6.1 (left) shows the chemical pilot plant VERA (german abbreviation for "**VER**fahrenstechnische Pilot**A**nlage") at the Institute of Automation and Computer Control. The plant is used for experimentally evaluating automation and control methods in a realistic, industrial environment. Therefore, standard industrial components are used throughout the plant [5, 104].

The main components of the plant are 8 tanks which can be flexibly connected by a complex pipe system. The tanks in the upper part of the figure are used to provide raw substances or store intermediate products, whereas the reactor tanks below are equipped with stirrers, heaters, coolers and various sensors to perform exothermic as well as endothermic reactions. Waste water is stored in the tank TW.

More than 70 sensors and 80 actuators are included in the plant. The discrete and continuous valves and pumps are used to control the medium flow in the plant, and the sensors allow to measure physical quantities such as level, volume flow, temperature, electrical conductivity and pH-value at locations of interest.

The plant automation is based on two industrial programmable logic controllers of type SIMATIC S7-300, which host a plant protection system and provide an interface to rapid prototyping software. Rapid control prototyping uses MATLAB/Simulink, while the process visualisation and manual operation are provided by WinCC software, each implemented on dedicated workstations [104].

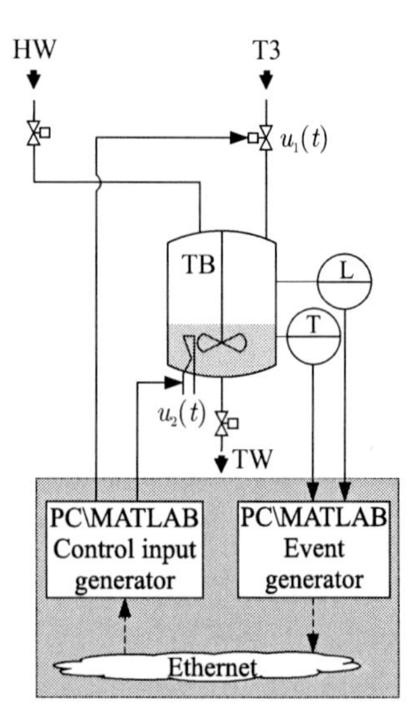

Figure 6.1.: Chemical pilot plant VERA (left) and implementations of the event-based state-feedback loop (right)

6.2. Implementation of the event-based state feedback

The plant provides two possibilities for implementing the event-based state-feedback loop (right-hand side of Fig 6.1):

1. The smart sensor (event generator) and the smart actuator (control input generator) can be realised on two separate workstations equipped with MATLAB/Simulink. These workstations are linked over a 100 Mbit/s Ethernet network.

2. The smart components can be realised on a single workstation (indicated by the dotted box). Here, the digital network between the components is also emulated by using MATLAB/Simulink.

Both implementations usually provide identical results as the 100 Mbit/s Ethernet guarantees ideal properties of the communication link in relation to the slow dynamics of the thermofluid process. Therefore, the second implementation has been generally preferred in the experiments because it allows not only a simpler installation but also a desired and reproducible adjustment of communication effects, which is utilised in Section 6.4. In both implementations, the control input generator and the event generator are periodically executed with the sampling time $T_s = 0.5$ s.

The controller matrices

$$\boldsymbol{K} = \begin{pmatrix} 0.08 & -0.18 \\ 0.17 & 0.74 \end{pmatrix} \tag{6.1}$$

for the proportional event-based controller and

$$\boldsymbol{K} = \begin{pmatrix} 0.13 & -0.18 & 0.0008 & -0.002 \\ 0.35 & 1.05 & 0.0018 & 0.008 \end{pmatrix}$$

for the event-based PI controller have been used in the experiments. They are slightly modified compared to the controllers (3.27), (4.33) applied in the simulation as they better meet the practical requirements.

The event generator alerts events, if

$$\|\boldsymbol{x}(t) - \boldsymbol{x}_s(t)\|_\infty \geq 2 \tag{6.2}$$

is detected. This inequality replaces the equality (3.21) due to the periodic operation of the event generator as discussed in Section 4.5. However, as the sampling is very fast in comparison to the time constants of the process, the consequences are marginal and can be neglected.

Note that in the following plots the actual level and the actual temperature of the medium in tank TB and not their deviations from the operating point

$$\bar{\boldsymbol{x}} = \begin{pmatrix} \bar{x}_1 \\ \bar{x}_2 \end{pmatrix} = \begin{pmatrix} \bar{l}_{\text{TB}} \\ \bar{\vartheta}_{\text{TB}} \end{pmatrix} = \begin{pmatrix} 40\,\text{cm} \\ 313\,\text{K} \end{pmatrix} \tag{6.3}$$

are indicated. This likewise holds for the input signal $\boldsymbol{u}(t)$ with

$$\bar{\boldsymbol{u}} = \begin{pmatrix} \bar{u}_1 \\ \bar{u}_2 \end{pmatrix} = \begin{pmatrix} 0.36 \\ 1.65 \end{pmatrix}. \tag{6.4}$$

Simulation vs. Experiment. In contrast to the simulation, the experimental set-up is affected by model uncertainties and measurement noise. The model uncertainties may result from a potential deviation of the actual state $\boldsymbol{x}(t)$ from the operating point $\bar{\boldsymbol{x}}$ used for the linearisation (Section 2.2) but are mainly due to the nonlinear characteristics of the continuous valves involved. Compared to model uncertainties and the influence of exogenous disturbances, the effect of measurement noise is negligible. Hence, as neither the capability of the smart components nor the communication network are subject to restrictions, the influence of model uncertainties on the behaviour of the event-based control, which has been investigated in Section 3.4.6, can be explicitly evaluated in the experiments.

Section 6.3 compares the experimental results obtained by applying the proportional event-based controller, the event-based PI controller and the augmented disturbance estimation under ideal communication properties. Section 6.4 shows experimental results obtained by artificially imposing communication imperfections, i.e. delays and quantisation.

6.3. Event-based control with ideal communication

6.3.1. Proportional event-based control

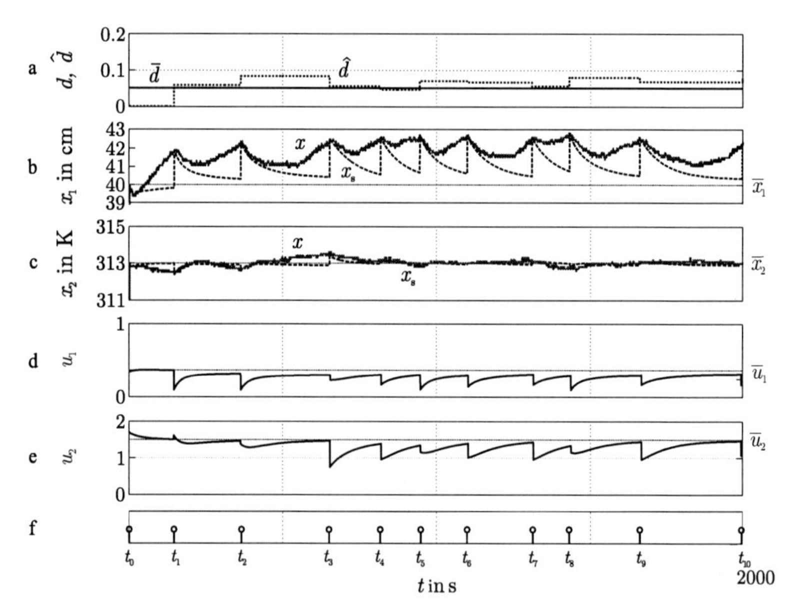

Figure 6.2.: Trajectories of the proportional event-based state-feedback loop. Solid lines: plant state $\boldsymbol{x}(t)$; dashed lines: model state $\boldsymbol{x}_s(t)$; dotted lines: operating point $\bar{\boldsymbol{x}}, \bar{\boldsymbol{u}}$.

Figure 6.2 shows the trajectories of the proportional event-based state-feedback loop subject to the constant disturbance $\bar{d} = 0.05$ (Fig. 6.2a).

The control aim is to stabilise the operating point $\bar{\boldsymbol{x}}$ given in Eq. (6.3). The behaviour of the level $x_1(t)$ and the behaviour of the fluid temperature $x_2(t)$ in reactor TB are depicted in Figs. 6.2b,c. An event takes place at time t_1, where inequality (6.2) is satisfied (Fig. 6.2b). At this time instance, the disturbance magnitude $\bar{d}$ is estimated (Fig. 6.2a). Overall 11 events occur in the time interval $[0, 2000\ \mathrm{s}]$ (Fig. 6.2f).

The experimental results differ slightly from the simulation results in Fig. 3.7 due to the model uncertainties mentioned, which affect the disturbance estimation and cause the variation of the disturbance estimate $\hat{d}_k$ at the event times (see Section 3.4.6). More concretely, these uncertainties lead to the divergence between $\boldsymbol{x}(t)$ and $\boldsymbol{x}_{\mathrm{s}}(t)$ after event time t_1 despite the good disturbance estimate $\hat{d}_1$.

Figures 6.2d and 6.2e show the control inputs $u_1(t)$ (valve angle) and $u_2(t)$ (heating power). Their discontinuous behaviour results from the update of the model state $\boldsymbol{x}_{\mathrm{s}}(t)$ at event times t_k which directly affects the control input according to $\boldsymbol{u}(t) = -\boldsymbol{K}\boldsymbol{x}_{\mathrm{s}}(t)$. The constant inputs $\bar{u}_1$ and $\bar{u}_2$ (Eq. (6.4)), which correspond to the operating point (6.3), are indicated as dotted lines.

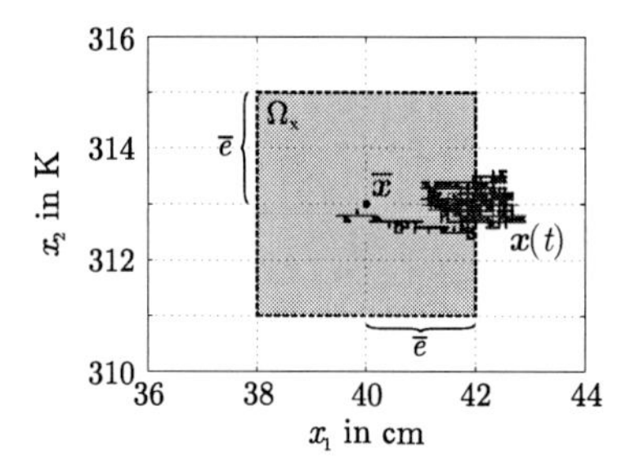

Figure 6.3.: Trajectories of the proportional event-based state-feedback loop in the state space

As in the simulation (left-hand side of Fig. 3.7, page 38), both $x_1(t)$ and $x_{\mathrm{s},1}(t)$ remain above the operating point, whereas the behaviour of the temperature is close to the operating point. Stationarily,

$$\boldsymbol{x}(t) \in \Omega_{\mathrm{x}}(\bar{\boldsymbol{x}}) = \{\boldsymbol{x} : \|\boldsymbol{x} - \bar{\boldsymbol{x}}\| \leq \bar{e}\} \tag{6.5}$$

can only be guaranteed in an approximate way as shown in Fig. 6.3.

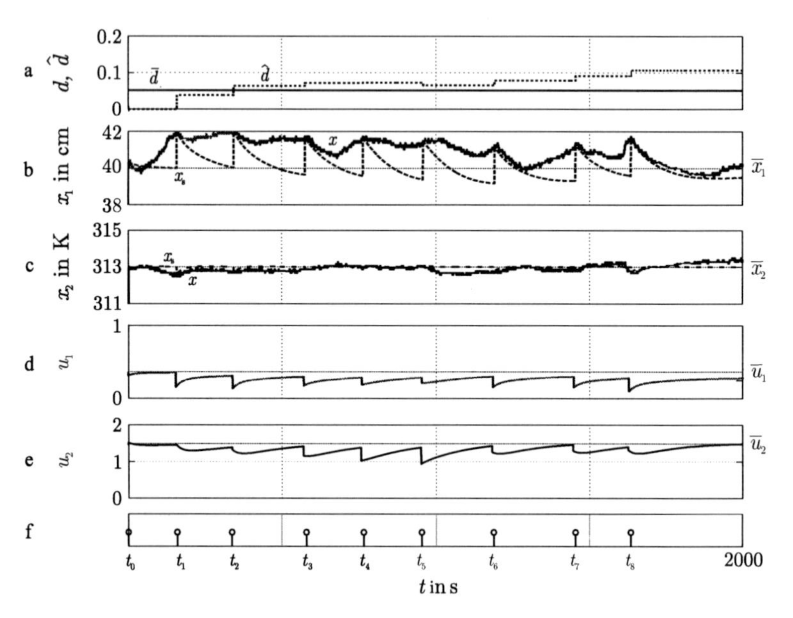

Figure 6.4.: Trajectories of the event-based PI-control loop. Solid lines: plant state $\boldsymbol{x}(t)$; dashed lines: model state $\boldsymbol{x}_{\mathrm{s}}(t)$; dotted lines: operating point $\bar{\boldsymbol{x}}$, $\bar{\boldsymbol{u}}$.

6.3.2. Event-based PI control

The trajectories of the event-based PI-control loop affected by the constant disturbance $\bar{d} = 0.05$ are depicted in Figure 6.4. An event takes place at time t_1 due to the level which exceeds the event condition (6.2) (Fig. 6.4b). Overall 9 events occur in the time interval $[0, 2000\ \mathrm{s}]$ (Fig. 6.4f) which shows a slightly reduced communication compared to the proportional event-based control. However, the influence of model uncertainties cannot be completely compensated which is indicated by the behaviour of $x_1(t)$ and $x_{\mathrm{s},1}(t)$ after the event time t_1.

In comparison to the experimental results discussed before, the trajectories show that the measured states $x_1(t)$ and $x_2(t)$ as well as the model states $x_{\mathrm{s},1}(t)$ and $x_{\mathrm{s},2}(t)$ stationarily move around the operating point $\bar{\boldsymbol{x}}$ and the relation (6.5) holds for all times t, which is depicted in Fig. 6.5. In summary, the behaviour of the event-based PI-control loop is considerably improved in comparison to the behaviour of the proportional event-based control loop. It better compensates the effect of model uncertainties.

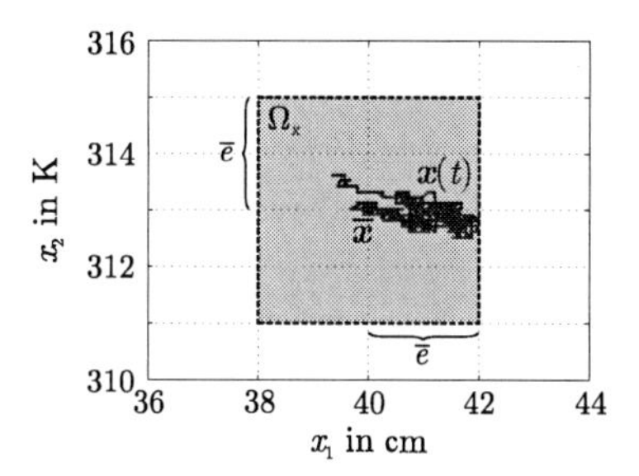

Figure 6.5.: Trajectories of the event-based PI-control loop in the state space

Setpoint change. Figure 6.6 presents further experimental results obtained by applying the event-based PI control. The event-based control loop is affected by the constant disturbance $\bar{d} = 0.05$ (Fig. 6.6a) and a change of the reference signal at time $t = 400$ s (Figs. 6.6b,c; dotted lines):

$$\boldsymbol{w}(t) = \begin{cases} \bar{\boldsymbol{x}} & \text{for } 0 \leq t < 400 \text{ s} \\ \bar{\boldsymbol{w}} = \begin{pmatrix} 43 \text{ cm} \\ 312 \text{ K} \end{pmatrix} & \text{for } t \geq 400 \text{ s}. \end{cases}$$

Overall 15 events occur in the time interval $[0, 2000 \text{ s}]$ (Fig. 6.6f) which shows an increased communication compared to the previous experiments. The reason is an increased approximation error of the nonlinear process dynamics by the linear model (2.7) used in the control input generator and the event generator because the setpoint $\bar{\boldsymbol{w}}$ increases the deviation from the operating point (6.3) around which the process was linearised.

Stationarily, the model states $x_{s,1}(t)$ and $x_{s,2}(t)$ move around the setpoint $\bar{\boldsymbol{w}}$ and, hence, the measured states $x_1(t)$ and $x_2(t)$ remain in a close surrounding as well.

Remarks. The experiments show that the event-based state feedback is robust against severe model uncertainties, because the difference of the model and the measured output is dealt with as disturbance, for which an estimate is used as an additional input to the control input generator. This can be seen by the varying disturbance estimates and the considerable deviation between the disturbance estimate and the actual disturbance.

In this context, Section 4.4.1 has proposed to augment the disturbance estimation in order to provide the event-based control more flexibility for dealing with model uncertainties, which is evaluated next.

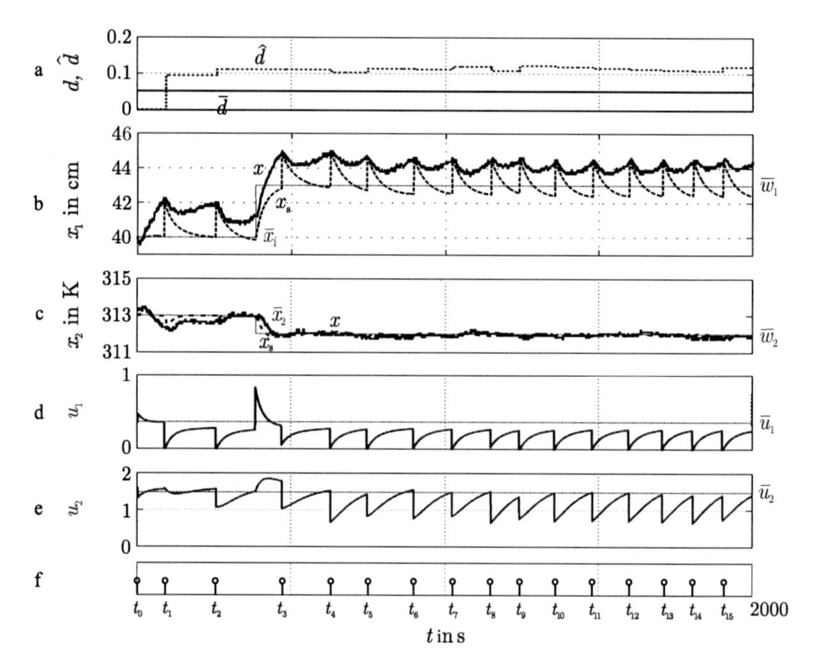

Figure 6.6.: Trajectories of the event-based PI-control loop with a change of the reference signal $\boldsymbol{w}(t)$. Solid lines: plant state $\boldsymbol{x}(t)$; dashed lines: model state $\boldsymbol{x}_s(t)$; dotted lines: operating point $\bar{\boldsymbol{x}}$, $\bar{\boldsymbol{u}}$.

6.3.3. Event-based PI control with augmented disturbance vector

Figure 6.7 shows the trajectories of the event-based PI-control loop which uses the augmented disturbance vector $\boldsymbol{d}_a(t)$ according to Section 4.4.1 in the disturbance estimation and the control input generation.

Instead of 8 events, which occur when using the original disturbance estimation (Fig. 6.4), only three events occur in the time interval $[0, 2000\text{ s}]$ with the augmented disturbance estimate. The reason is that the disturbance estimator has now the flexibility to independently estimate the influence of the disturbance and the model uncertainties on each single state variable.

After event time t_2, the actual level behaviour $x_1(t)$ and the model behaviour $x_{s,1}(t)$ are nearly identical but the difference in the temperature behaviour is increased compared to the results shown in Fig. 6.4. However, this difference remains inside the tolerance bound speci-

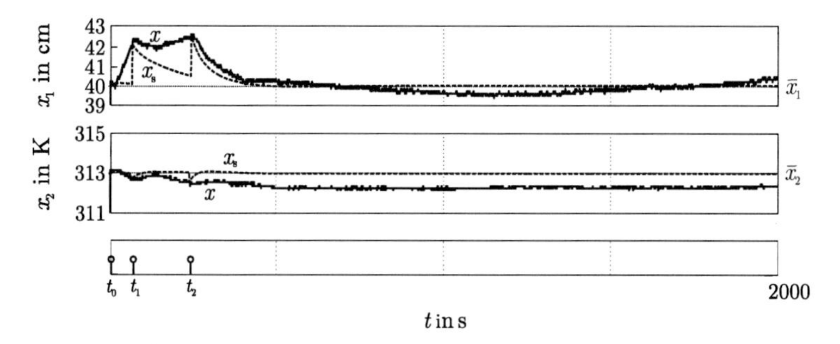

Figure 6.7.: Trajectories of the event-based PI-control loop with the augmented disturbance vector $\boldsymbol{d}_{\mathrm{a}}(t)$. Solid lines: plant state $\boldsymbol{x}(t)$; dashed lines: model state $\boldsymbol{x}_{\mathrm{s}}(t)$; dotted lines: operating point $\bar{\boldsymbol{x}}$.

fied according to inequality (6.2) so that no further event occurs. As the stationary deviation results from the event-based character of the control loop, it can be expected for any implementation of event-based control.

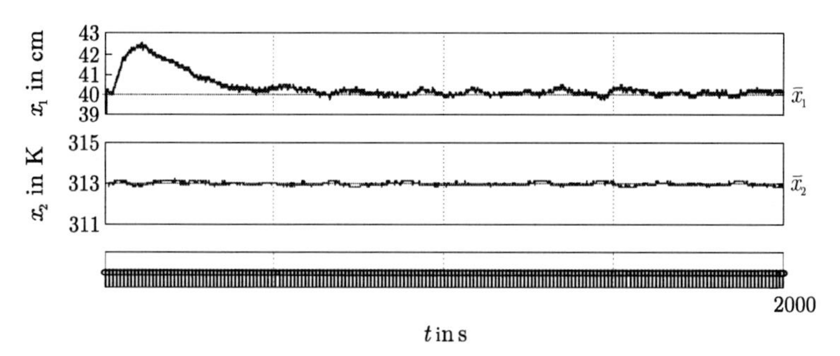

Figure 6.8.: Trajectories of the discrete-time PI-control loop with $T_{\mathrm{s}} = 10$ s

For comparison, the trajectories of the discrete-time PI-control loop with the sampling period $T_{\mathrm{s}} = 10$ s are depicted in Fig. 6.8. It can be seen that both state variables $x_1(t)$ and $x_2(t)$ reach the operating point but the overall performance is similar to the behaviour of the event-based PI control. A more evident distinction is given by the communication exchange, where now 200 data transmissions occur instead of three communication events in the event-based case in the same time interval.

Thus, the event-based control is able to perform a comparable behaviour while significantly reducing the communication compared to conventional discrete-time control.

6.4. Event-based control subject to communication imperfections

The previous experiments have been executed with ideal communication properties to concentrate the investigations on the effect of model uncertainties. In the following experiments, communication imperfections, i.e. delays and quantisation, are considered. The communication link has been emulated by MATLAB/Simulink according to the second implementation discussed in Section 6.2 in order to arbitrarily and reproducibly adjust these communication effects.

6.4.1. Communication delays

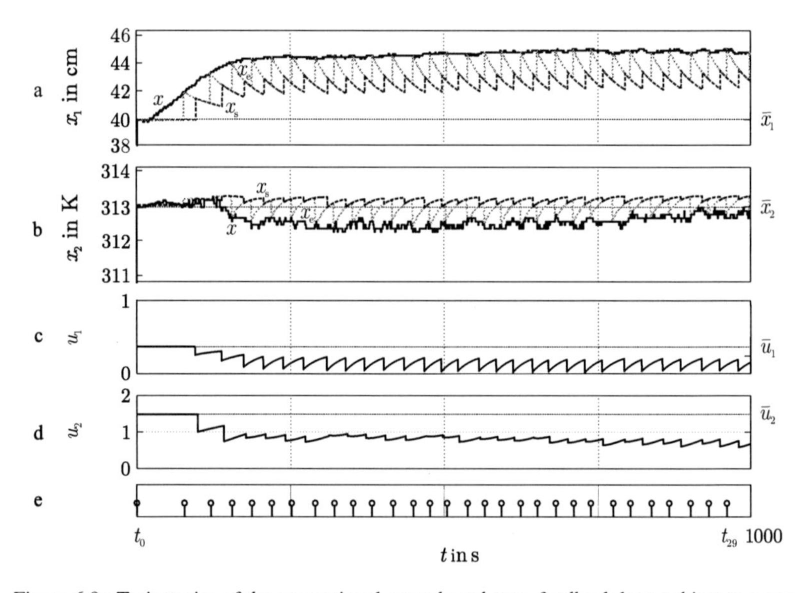

Figure 6.9.: Trajectories of the proportional event-based state-feedback loop subject to a constant delay. Solid lines: plant state $\boldsymbol{x}(t)$; dashed lines: model state $\boldsymbol{x}_\mathrm{s}(t)$; grey dotted lines: model state $\boldsymbol{x}_\mathrm{e}(t)$; dotted lines: operating point $\bar{\boldsymbol{x}}$, $\bar{\boldsymbol{u}}$.

The trajectories of the proportional event-based state-feedback loop subject to the constant disturbance $\bar{d} = 0.05$ and a constant communication delay are depicted in Fig. 6.9. To cope with model uncertainties and to avoid violating assumption (5.6) ($\tau_k < t_{k+1} - t_k,\ \forall k$, Section 5.2), the communication delay has been reduced to $\tau = 20$ s instead of $\tau = 77$ s in the simulation (Fig. 5.4, page 116). Moreover, no disturbance estimation has been applied.

In comparison to the simulation results, the model uncertainties lead to a much higher communication with overall 30 events in 1000 s even though the delay is significantly smaller. Nevertheless, the event-based control scheme works properly since the behaviour remains stable despite the delay.

6.4.2. Quantisation

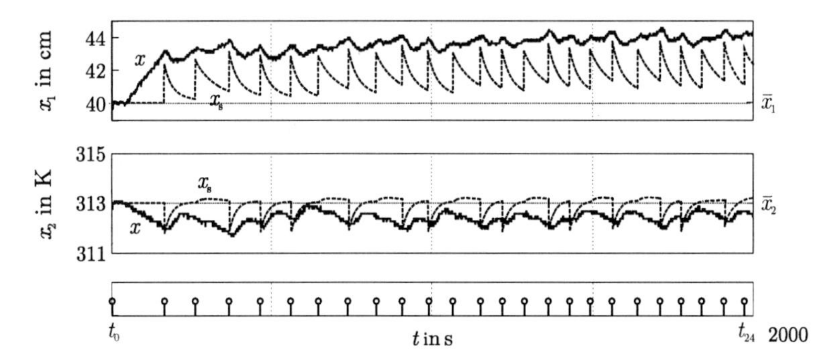

Figure 6.10.: Trajectories of the proportional event-based state-feedback loop with quantised communication. Solid lines: plant state $\boldsymbol{x}(t)$; dashed lines: model state $\boldsymbol{x}_s(t)$; dotted lines: operating point $\bar{\boldsymbol{x}}$.

Figure 6.10 shows the trajectories of the quantised event-based state-feedback loop affected by the constant disturbance $\bar{d} = 0.05$. The event threshold has been set to $\bar{e} = 3$ and the quantisation level is $\xi = 5$. Again, no disturbance estimation has been applied.

It can be seen that the concept works properly in the experimental environment as the deviation between the actual state $\boldsymbol{x}(t)$ and the operating point $\bar{\boldsymbol{x}}$ is only slightly increased compared to the simulation (Fig. 5.13, page 137) as well as the ideal communication case (Fig. 6.2), which results in both cases from the larger event threshold.

Nevertheless, it shows that model uncertainties mainly affect the communication (overall 25 events in 2000 s) since the communication is similar to the simulation (Fig. 5.13) even though the threshold is larger and the disturbance magnitude is much smaller. Moreover, the more frequent communication likewise results from the quantisation (cf. Section 5.4.6) which increases the information exchange in comparison to the ideal communication case (Fig. 6.2).

The quantised event-based control loop with incorporated disturbance estimation has been experimentally evaluated in [23].

Summary. The experiments clearly demonstrate the following properties of the event-based state feedback:

- It is robust against severe model uncertainties.
- It significantly reduces the communication in comparison to conventional discrete-time control.
- Model uncertainties and communication imperfections rather affect the information exchange than the performance of the event-based state-feedback loop.

Moreover, the event-based state-feedback control has been successfully evaluated in further experimental environments such as an inverted pendulum [17] or a twin-roter helicopter model [18].

7. Summary and outlook

7.1. Contribution of this thesis

This thesis has presented a state-feedback approach to event-based control which is capable of approximating the behaviour of a continuous-time state-feedback loop with arbitrary precision while adapting the communication to the effect of unknown disturbances.

The event-based state-feedback loop includes a control input generator and an event generator both of which use a model of the continuous-time state-feedback loop for the control input generation and the event generation. An event is generated whenever the difference between the behaviour of the event-based control loop and the behaviour of the reference model used in the event generator reaches a predefined event threshold $\bar{e}$. By choosing this threshold, a tradeoff between the desired performance and admissible communication properties of the event-based state-feedback loop has to be found.

Between two consecutive event times, the unknown disturbance is approximated by a disturbance estimator which provides the control input generator and the event generator with a disturbance estimate. This disturbance information is used by the control input generator to determine the control input in a feedforward way by means of the model of the continuous-time state-feedback loop.

This thesis has complemented the event-based state feedback by proposing

- suitable strategies for compensating imperfect plant information such as model uncertainties and non-measurable states (Sections 3.4.6, 4.2, 4.4, Theorems 7, 9), and
- modifications of the components to overcome imperfect properties of the communication link such as delays, packet losses and data-rate constraints (Chapter 5, Theorems 15, 16).

In all these cases, it has been shown that the event-based state-feedback loop remains stable despite the imperfections and a bound on the approximation error with respect to the continuous-time state-feedback loop can be specified. However, the performance of the event-based state-feedback loop in terms of the guaranteed approximation accuracy may deteriorate and an increased but bounded information exchange over the feedback link in the control loop is to be expected (Theorems 10, 17, 18).

Moreover, the event-based state feedback has been extended in Chapter 4 by

- incorporating a dynamical controller in the control input generator to improve the set-point tracking properties of the event-based state-feedback loop (Section 4.3, Theorem 12),
- conceptional improvements of the disturbance estimation (Section 4.4), and
- a discrete-time realisation of the control input generator and the event generator (Section 4.5, Theorems 13, 14).

Additionally, this thesis provides a Lyapunov-based stability analysis (Section 3.4.2, Theorem 4) and a common basis for the comparison of event-based state feedback and discrete-time state feedback (Chapter 3, Theorem 8).

Throughout the thesis, a thermofluid process has been used to illustrate the elaborated concepts and to evaluate the theoretical results. The experimental evaluation of the event-based controlled thermofluid process, which has been realised at a chemical pilot plant, has shown that the proposed concepts are robust against severe model uncertainties and that event-based control is able to significantly reduce the communication over the feedback link compared to conventional discrete-time control (Chapter 6).

7.2. Open problems

There are two primary directions for future research. First, remaining problems concern single-loop systems as considered in this thesis, in which the sensor node and the actuator node are assumed to have access to all measurements and actuators of the plant. Second, the event-based state-feedback approach can be extended to multi-loop systems, in which several control loops share a common communication medium.

Single-loop systems. The work in this thesis has been focussed on linear systems and, hence, an extension to nonlinear systems is nearby in order to enhance the practical relevance of the scheme. An extension to input-output linearisable systems has been already studied in [110, 111]. Further promising classes of nonlinear systems are Hammerstein and Wiener systems which possess static nonlinearities in the input or the output, respectively. A suitable approach to deal with such systems in the continuous-time case is given by considering a piecewise linear analysis [69] which may be adapted to the event-based scenario.

In literature, there are some approaches which successfully applied event-based control in a stochastic setting, e.g. [31, 61]. To improve the comparison between event-based control and discrete-time control in terms of reducing the conservatism, a stochastic analysis can be also applied to the approach presented in this thesis instead of the performed worst-case analysis.

A practical drawback of the event-based state feedback is that it requires a continuous or, at least, a periodic monitoring of the event condition and a demanding computational effort on the smart components. However, these components may underlie restrictions with respect to their computational resources and the energy consumption. In this case, the event-based state feedback should be realised in a self-triggered way [25, 126], which has been studied in [21].

The consideration of a self-triggered implementation reveals a problem, which would occur in the proposed event-based output-feedback control, as the incorporated state observer requires a continuous-time measurement. To circumvent this drawback, an event-based state observer should be designed based on [28, 77, 107, 119]. An event-based implementation of the state-set observer investigated in [22, 101, 102] may provide a promising alternative.

Multi-loop systems. The central application field for event-based control is given by networked control systems, in which the sensor and actuator nodes are usually widely distributed within the plant. Therefore, the assumption that the sensor node and the actuator node have access to all measurements and actuators of the plant is not feasible. Moreover, the reduction of the communication by event-based control becomes increasingly important if several control loops share a common communication medium (Fig. 1.2).

In these situations, a distributed realisation of the event-based state feedback is required [88, 103, 127, 129] which necessitates a new analysis or, at least, a significant adaptation of the current methods. However, the investigation of the behaviour between two consecutive event times, as carried out in this thesis, is inappropriate because the events are invoked asynchronously by each single control loop. Instead, the investigation of *common Lyapunov functions*, which have been used e.g. in the stability analysis of hybrid systems [2, 52, 85], may provide a suitable method to circumvent this problem. Reference [95] elaborated first concepts for realising the event-based state feedback of physically interconnected subsystems in a distributed way.

Multi-loop systems offer another challenge because the properties of the communication medium and the control have to be indispensably investigated with respect to their interaction. The consequences of various communication structures, communication protocols, and communication delays and packet losses on distributed event-based control or vice versa have been analysed e.g. in [36, 37, 40, 127] based on stochastic or Lyapunov-based methods. Similar approaches can be also applied for the event-based state-feedback control.

Bibliography

Contributions by the author:

Publications:

[1] L. Grüne, S. Jerg, O. Junge, D. Lehmann, J. Lunze, F. Müller, and M. Post. Two complementary approaches to event-based control. *Automatisierungstechnik*, 58(4): 173–182, 2010.

[2] W. P. M. H. Heemels, D. Lehmann, J. Lunze, and B. De Schutter. Introduction to hybrid systems. In J. Lunze and F. Lamnabhi-Lagarrigue, editors, *Handbook of Hybrid Systems Control*, pages 4–30. Cambridge University Press, 2009.

[3] D. Lehmann and J. Lunze. Event-based control: a state-feedback approach. In *Proceedings of European Control Conference*, pages 1716–1721, Budapest, 2009.

[4] D. Lehmann and J. Lunze. Event-based control using quantized state information. In *Proceedings of Workshop on Distributed Estimation and Control in Networked Systems*, Annecy, 2010.

[5] D. Lehmann and J. Lunze. Extension and experimental evaluation of an event-based state-feedback approach. *Control Engineering Practice*, 19(2):101–112, 2011.

[6] D. Lehmann and J. Lunze. Event-based control with communication delays. In *Proceedings of IFAC World Congress*, Milano, 2011 (to appear).

[7] D. Lehmann and J. Lunze. Event-based output-feedback control. In *Proceedings of Mediterranean Conference on Control and Automation*, Corfu, 2011.

[8] J. Lunze and D. Lehmann. A state-feedback approach to event-based control. *Automatica*, 46(1):211–215, 2010.

Technical reports:

[9] L. Grüne, S. Jerg, O. Junge, D. Lehmann, J. Lunze, F. Müller, and M. Post. Two complementary approaches to event-based control. Technical report, Ruhr-Universität Bochum, Lehrstuhl für Automatisierungstechnik und Prozessinformatik, 2010.

[10] D. Lehmann. Event-based control - Literature review and new approaches. Technical report, Ruhr-Universität Bochum, Lehrstuhl für Automatisierungstechnik und Prozessinformatik, 2008.

[11] D. Lehmann. A state-feedback approach to event-based control - Practical oriented extensions and experimental evaluation. Technical report, Ruhr-Universität Bochum, Lehrstuhl für Automatisierungstechnik und Prozessinformatik, 2009.

[12] D. Lehmann. Event-based control subject to communication delays. Technical report, Ruhr-Universität Bochum, Lehrstuhl für Automatisierungstechnik und Prozessinformatik, 2011.

[13] D. Lehmann. Analogies and distinctions between discrete-time control and event-triggered control. Technical report, Ruhr-Universität Bochum, Lehrstuhl für Automatisierungstechnik und Prozessinformatik, 2011.

[14] D. Lehmann and J. Pfahler. Füllstands- und Leitfähigkeitsregelung an der verfahrenstechnischen Pilotanlage VERA. Technical report, Ruhr-Universität Bochum, Lehrstuhl für Automatisierungstechnik und Prozessinformatik, 2009.

Supervised theses:

[15] T. Arndt. Vergleich von verschiedenen ereignisbasierten Regelungskonzepten. Study thesis, Ruhr-Universität Bochum, Lehrstuhl für Automatisierungstechnik und Prozessinformatik, 2008.

[16] M. Caspar. Event-based realisation of an adaptive control scheme. Diploma thesis, Ruhr-Universität Bochum, Lehrstuhl für Automatisierungstechnik und Prozessinformatik, 2010.

[17] K. Freynik. Ereignisbasierte Regelung eines invertierten Pendels. Study thesis, Ruhr-Universität Bochum, Lehrstuhl für Automatisierungstechnik und Prozessinformatik, 2009.

[18] J. Z. Iglesias. Event-based control of a twin-rotor system. Study project, Ruhr-Universität Bochum, Lehrstuhl für Automatisierungstechnik und Prozessinformatik, 2010.

[19] P. Otto. Erweiterung eines Konzeptes zur ereignisbasierten Regelung. Study thesis, Ruhr-Universität Bochum, Lehrstuhl für Automatisierungstechnik und Prozessinformatik, 2009.

[20] K. Schlüter. Untersuchung des Einflusses von Modellunsicherheiten auf die ereignisbasierte Regelung. Study thesis, Ruhr-Universität Bochum, Lehrstuhl für Automatisierungstechnik und Prozessinformatik, 2009.

[21] F. Stadtmann. Selbstgetriggerte Realisierung der ereignisbasierten Zustandsrückführung. Study thesis, Ruhr-Universität Bochum, Lehrstuhl für Automatisierungstechnik und Prozessinformatik, 2011.

[22] M. Wegener. Ereignisbasierte Regelung bei nichtmessbarem Zustand. Diploma thesis, Ruhr-Universität Bochum, Lehrstuhl für Automatisierungstechnik und Prozessinformatik, 2009.

[23] P. Weizinger. Ereignisbasierte Regelung mit quantisierter Zustandsinformation. Study thesis, Ruhr-Universität Bochum, Lehrstuhl für Automatisierungstechnik und Prozessinformatik, 2010.

Further literature:

[24] J. Ackermann. *Robust Control.* Springer-Verlag, London, 2002.

[25] A. Anta and P. Tabuada. To sample or not to sample: self-triggered control for nonlinear systems. *IEEE Transactions on Automatic Control*, 55(9):2030–2042, 2010.

[26] A. Anta and P. Tabuada. On the minimum attention and anytime attention problems for nonlinear systems. In *Proceedings of IEEE Conference on Decision and Control*, pages 3234–3239, Atlanta, 2010.

[27] K.-E. Årzén. A simple event-based PID controller. In *Proceedings of IFAC World Congress*, pages 423–428, Beijing, 1999.

[28] K. J. Åström. Event based control. In A. Astolfi and L. Marconi, editors, *Analysis and Design of Nonlinear Control Systems*, pages 127–147. Springer-Verlag, Berlin, 2008.

[29] K. J. Åström. *Introduction to Stochastic Control Theory*. Academic Press, New York, 1970.

[30] K. J. Åström and B. Bernhardsson. Systems with Lebesgue sampling. In *Directions in Mathematical Systems Theory and Optimization*, volume 286 of *Lecture Notes in Control and Information Sciences*. Springer-Verlag, Berlin Heidelberg, 2003.

[31] K. J. Åström and B. Bernhardsson. Comparison of Riemann and Lebesgue sampling for first order stochastic systems. In *Proceedings of IEEE Conference on Decision and Control*, volume 2, pages 2011–2016, Las Vegas, 2002.

[32] K. J. Åström and B. Wittenmark. *Computer-Controlled Systems*. Prentice Hall, Englewood Cliffs, 1990.

[33] K. J. Åström and B. Wittenmark. *Adaptive Control*. Addison-Wesley, 1994.

[34] A. Bemporad, W. P. M. H. Heemels, and M. Johansson. *Networked Control Systems*. Lecture Notes in Control and Information Sciences. Springer-Verlag, London, 2010.

[35] D. S. Bernstein. *Matrix Mathematics*. Princeton University Press, 2005.

[36] R. Blind and F. Allgöwer. Analysis of networked event-based control with a shared communication medium: part I - pure ALOHA. In *Proceedings of IFAC World Congress*, Milano, 2011 (to appear).

[37] R. Blind and F. Allgöwer. Analysis of networked event-based control with a shared communication medium: part II - slotted ALOHA. In *Proceedings of IFAC World Congress*, Milano, 2011 (to appear).

[38] S. Boyd, L. El. Ghaoui, E. Feron, and V. Balakrishnan. *Linear Matrix Inequalities in System and Control Theory*. SIAM, 1994.

[39] R. W. Brockett and D. Liberzon. Quantized feedback stabilization of linear systems. *IEEE Transactions on Automatic Control*, 47(7):1279–1289, 2000.

[40] A. Cervin and T. Henningsson. Scheduling of event-triggered controllers on a shared network. In *Proceedings of IEEE Conference on Decision and Control*, pages 3601–3606, Cancun, 2008.

[41] D. Chakraborty. *Need-based feedback: An optimization approach*. PhD thesis, University of Florida, 2007.

[42] R. Cogill. Event-based control using quadratic approximate value functions. In *Proceedings of IEEE Conference on Decision and Control*, pages 5883–5888, Shanghai, 2009.

[43] D. de Bruin and P. P. J. van den Bosch. Measurement of the lateral vehicle position with permanent magnets. In *Proceedings of IFAC Workshop on Intelligent Components for Vehicles*, pages 9–14, Seville, 1998.

[44] D. V. Dimarogonas and K. H. Johansson. Event-triggered control for multi-agent systems. In *Proceedings of IEEE Conference on Decision and Control*, pages 7131–7136, Shanghai, 2009.

[45] D. V. Dimarogonas, E. Frazzoli, and K. H. Johansson. Distributed self-triggered control for multi-agent systems. In *Proceedings of IEEE Conference on Decision and Control*, pages 6716–6721, Atlanta, 2010.

[46] M. C. F. Donkers and W. P. M. H. Heemels. Output-based event-triggered control with guaranteed L_2-gain and improved event-triggering. In *IEEE Conference on Decision and Control*, pages 3246–3251, Atlanta, 2010.

[47] A. Eqtami, D. V. Dimarogonas, and K. J. Kyriakopoulos. Event-triggered strategies for decentralized model predictive controllers. In *Proceedings of IFAC World Congress*, Milano, 2011 (to appear).

[48] L. Evans. *The Large Hadron Collider: A Marvel of Technology*. CRC Press, 2009.

[49] F. Fagnani and S. Zampieri. Stability analysis and synthesis for scalar linear systems with a quantized feedback. *IEEE Transactions on Automatic Control*, 48(9):1569–1584, 2003.

[50] J. Falkenhain. Operatornormen für den Reglerentwurf. Technical report, Ruhr-Universität Bochum, Lehrstuhl für Automatisierungstechnik und Prozessinformatik, 2008.

[51] H. Gao and T. Chen. A new approach to quantized feedback control systems. *Automatica*, 44(2):534–542, 2008.

[52] R. Goebel, R. G. Sanfelice, and A. R. Teel. Hybrid dynamical systems. *IEEE Control Systems Magazine*, 29(2):28–93, 2009.

[53] L. Grüne and F. Müller. An algorithm for event-based optimal feedback control. In *Proceedings of IEEE Conference on Decision and Control*, pages 5311–5316, Shanghai, 2009.

[54] W. M. Haddad, V. Chellaboina, and S. G. Nersesov. *Impulsive and Hybrid Dynamical Systems: Stability, Dissipativity, and Control.* Princeton University Press, 2006.

[55] W. P. M. H. Heemels, R. J. A. Gorter, A. van Zijl, P. P. J. van den Bosch, S. Weiland, W. H. A. Hendrix, , and M. R. Vonder. Asynchronous measurement and control: a case study on motor synchronization. *Control Engineering Practice*, 7(12):1467–1482, 1999.

[56] W. P. M. H. Heemels, J. Sandee, and P. van den Bosch. Analysis of event-driven controllers for linear systems. *International Journal of Control*, 81(4):571–590, 2008.

[57] K. auf der Heide, K. Janschek, and A. Tkocz. Synergy in power and momentum management for spacecraft using double gimballed solar arrays. In *Proceeding of IFAC Symposium on Automatic Control in Aerospace*, pages 290–295, St. Petersburg, 2004.

[58] T. Henningsson. *Event-based control and estimation with stochastic disturbances*. PhD thesis, Lund University, Department of Automatic Control, 2008.

[59] T. Henningsson and A. Cervin. Event-based control over networks: Some research questions and preliminary results. Technical report, Lund University, Department of Automatic Control, 2006.

[60] T. Henningsson and A. Cervin. Comparison of LTI and event-based control for a moving cart with quantized position measurements. In *Proceedings of European Control Conference*, pages 3791–3796, Budapest, 2009.

[61] T. Henningsson, E. Johannesson, and A. Cervin. Sporadic event-based control of first-order linear stochastic systems. *Automatica*, 44(11):2890–2895, 2008.

[62] J. Hespanha, P. Naghshtabrizi, and Y. Xu. A survey of recent results in networked control systems. *Proceedings of the IEEE*, 95(1):138–162, 2007.

[63] S. Hirche, M. Buss, P. Hinterseer, and E. Steinbach. Towards deadband control in networked teleoperation systems. In *Proceedings of IFAC World Congress*, Prague, 2005.

[64] K. C. Howell and H. J. Pernickat. Stationkeeping method for libration point trajectories. *Journal of Guidance, Control, and Dynamics*, 16(1):151–159, 1993.

[65] D. Hristu-Varsakelis and P. R. Kumar. Interrupt-based feedback control over a shared communication medium. In *Proceedings of IEEE Conference on Decision and Control*, volume 3, pages 3223–3228, Las Vegas, 2002.

[66] D. Hristu-Varsakelis and W. S. Levine. *Handbook of Networked and Embedded Control Systems*. Birkhäuser Verlag, Boston, 2005.

[67] P. A. Ioannou and J. Sun. *Robust Adaptive Control*. Prentice Hall, 1996.

[68] A. Isidori. *Nonlinear Control Systems*. Springer-Verlag, New York, 1995.

[69] M. Johansson. *Piecewise linear control systems*, volume 284 of *Lecture notes in control and information sciences*. Springer-Verlag, Heidelberg, 2002.

[70] H. K. Khalil. *Nonlinear Systems*. Prentice Hall, New Jersey, 2002.

[71] N. V. Kirianaki, S. Y. Yurish, N. O. Shpak, and V. O. Deynega. *Data Acquisition and Signal Processing for Smart Sensors*. Wiley, West Sussex, 2002.

[72] E. Kofman and J. H. Braslavsky. Level crossing sampling in feedback stabilization under data-rate constraints. In *Proceedings of IEEE Conference on Decision and Control*, pages 4423–4428, San Diego, 2006.

[73] V. Lakshmikantham, S. Leela, and A. A. Martynyuk. *Practical stability of nonlinear systems*. World Scientific, 1990.

[74] M. D. Lemmon. Event-triggered feedback in control, estimation, and optimization. In A. Bemporad, W. P. M. H. Heemels, and M. Johansson, editors, *Networked Control Systems*, volume 405 of *Lecture Notes in Control and Information Sciences*, pages 293–358. Springer-Verlag, Berlin Heidelberg, 2010.

[75] P Levis, S. Madden, D. Gay, J. Polastre, R. Szewczyk, A. W. E. Brewer, and D. Culler. The emergence of networking abstractions and techniques in TinyOS. In *Proceedings of USENIX/ACM Symposium on Networked Systems Design and Implementation*, pages 1–14, San Francisco, 2004.

[76] D. Liberzon. On stabilization of linear systems with limited information. *IEEE Transactions on Automatic Control*, 48(2):304–307, 2003.

[77] L. Lichun and M. Lemmon. Event-triggered output feedback control of finite horizon discrete-time multi-dimensional linear processes. In *Proceedings of IEEE Conference on Decision and Control*, pages 3221–3226, Atlanta, 2010.

[78] J. Lunze. Diagnosis of quantized systems based on a timed discrete-event model. *IEEE Transactions on Systems, Man and Cybernetics, Part A*, 30(3):322–335, 2000.

[79] J. Lunze. *Regelungstechnik 1*. Springer-Verlag, Heidelberg, 2005.

[80] J. Lunze. *Automatisierungstechnik*. Oldenbourg Verlag, München, 2008.

[81] J. Lunze. Event-based control: a state-feedback approach. Technical report, Ruhr-Universität Bochum, Lehrstuhl für Automatisierungstechnik und Prozessinformatik, 2008.

[82] J. Lunze. Event-based control with delayed information exchange. Technical report, Ruhr-Universität Bochum, Lehrstuhl für Automatisierungstechnik und Prozessinformatik, 2009.

[83] J. Lunze. *Regelungstechnik 2*. Springer-Verlag, Berlin Heidelberg, 2010.

[84] J. Lunze. Event-based control: a tutorial introduction. *Journal of the Society of Instrument and Control Engineers*, 49(11):783–788, 2010.

[85] J. Lunze and F. Lamnabhi-Lagarrigue. *Handbook of Hybrid Systems Control*. Cambridge University Press, 2009.

[86] M. Mazo and P. Tabuada. Input-to-state stability of self-triggered control systems. In *Proceedings of IEEE Conference on Decision and Control*, pages 928–933, Shanghai, 2009.

[87] M. Mazo and P. Tabuada. Towards decentralized event-triggered implementations of centralized control laws. In *Proceedings of International Workshop on Networks of Cooperating Objects*, Stockholm, 2010.

[88] M. Mazo and P. Tabuada. Event-triggered and self-triggered control over sensor/ actuator networks. In *Proceedings of IEEE Conference on Decision and Control*, pages 435–440, Cancun, 2008.

[89] M. Mazo, A. Anta, and P. Tabuada. On self-triggered control for linear systems: guarantees and complexity. In *Proceedings of European Control Conference*, pages 3767–3772, Budapest, 2009.

[90] A. Molin and S. Hirche. Structural characterization of optimal event-based controllers for linear stochastic systems. In *Proceedings of IEEE Conference on Decision and Control*, pages 3227–3233, Atlanta, 2010.

[91] L. A. Montestruque. *Model-based networked control systems.* PhD thesis, University of Notre Dame, Graduate School, 2004.

[92] L. A. Montestruque and P. J. Antsaklis. On the model-based control of networked systems. *Automatica*, 39(10):1837–1843, 2003.

[93] G. N. Nair, F. Fagnani, S. Zampieri, and R. J. Evans. Feedback control under data rate constraints: an overview. *Proceeding of the IEEE*, 95(1):108–137, 2007.

[94] M. Nechyba and Y. Xu. On discontinuous human control strategies. In *Proceedings of IEEE International Conference on Robotics and Automation*, pages 2237–2243, Leuven, 1998.

[95] H. C. N. Nguyen. Ereignisbasierte Regelung gekoppelter Systeme. Diploma thesis, Ruhr-Universität Bochum, Lehrstuhl für Automatisierungstechnik und Prozessinformatik, 2011.

[96] K. Ogata. *Discrete-Time Control Systems.* Prentice Hall, 1995.

[97] P. G. Otanez, J. G. Moyne, and D. M. Tilbury. Using deadbands to reduce communication in networked control systems. In *Proceedings of American Control Conference*, pages 3015–3020, Anchorage, 2002.

[98] F. Palutan, D. De Matino, S. Falzini, and M. Melis. Geostationary station keeping by ion thrusters: genetic algorithms and optimization. *International Journal of satellite communications and networking*, 14(1):1–9, 1996.

[99] C. De Persis. Robust stabilization of nonlinear systems by quantized and ternary control. In *Proceedings of IFAC World Congress*, pages 5191–5196, Seoul, 2008.

[100] C. De Persis and A. Isidori. Stabilizability by state feedback implies stabilizability by encoded state feedback. *Systems & Control Letters*, 53(3-4):249–258, 2004.

[101] P. Planchon. *Guaranteed diagnosis of uncertain linear systems using state-set observation.* Logos-Verlag, Berlin, 2007.

[102] P. Planchon and J. Lunze. Diagnosis of linear systems with structured uncertainties based on guaranteed state observation. *International Journal of Control, Automation and Systems*, 6(3):308–319, 2008.

[103] M. Rabi and K. H. Johansson. Scheduling packets for event-triggered control. In *Proceedings of European Control Conference*, pages 3779–3784, Budapest, 2009.

[104] J. H. Richter, T. Schlage, and J. Lunze. Control reconfiguration of a thermofluid process by means of a virtual actuator. *IET Proceedings on Control Theory and Applications*, 1 (6):1606–1620, 2007.

[105] J. Sandee. *Event-driven control in theory and practice*. PhD thesis, Technische Universiteit Eindhoven, 2006.

[106] G. S. Seyboth, D. V. Dimarogonas, and K. H. Johansson. Control of multi-agent systems via event-based communication. In *Proceedings of IFAC World Congress*, Milano, 2011 (to appear).

[107] J. Sijs, M. Lazar, and W. P. M. H. Heemels. On integration of event-based estimation and robust MPC in a feedback loop. In *Hybrid Systems: Computation and control*, pages 31–40, Stockholm, 2010.

[108] R. Sipahi, S.-I. Niculescu, C. T. Abdallah, W. Michels, and K. Gu. Stability and stabilization of systems with time delay. *IEEE Control Systems Magazine*, 31(1):38–65, 2011.

[109] E. D. Sontag. Input to state stability: basic concepts and results. In P. Nistri and G. Stefani, editors, *Nonlinear and Optimal Control Theory*, pages 163–220. Springer-Verlag, Berlin, 2006.

[110] C. Stöcker and J. Lunze. Event-based control of input-output linearizable systems. In *Proceedings of IFAC World Congress*, Milano, 2011 (to appear).

[111] C. Stöcker and J. Lunze. Event-based control of nonlinear systems: an input-output linearization approach. In *Proceedings of IEEE Conference on Decision and Control*, Orlando, 2011 (submitted).

[112] T. Steffen, L. Litz, A. Bauchspieß, and J. Y. Ishihara. Stable sampling period adaptation for energy saving in ambient intelligence networks. In *Congresso Brasileiro de Automática*, pages 4469–4475, Bonito, 2010.

[113] P. Tabuada. Event-triggered real-time scheduling of stabilizing control tasks. *IEEE Transactions on Automatic Control*, 52(9):1680–1685, 2007.

[114] P. Tabuada and X. Wang. Preliminary results on state-triggered scheduling of stabilizing control tasks. In *Proceedings of IEEE Conference on Decision and Control*, pages 282–287, San Diego, 2006.

[115] A. S. Tanenbaum. *Computernetzwerke*. Pearson Studium, Munich, 2003.

[116] U. Tiberi, C. Fischione, K. H. Johansson, and M. D. Di Benedetto. Adaptive self-triggered control over IEEE 802.15.4 networks. In *Proceedings of IEEE Conference on Decision and Control*, pages 2099–2104, Atlanta, 2010.

[117] Y. Tipsuwan and M. Y. Chow. Control methodologies in networked control systems. *Control Engineering Practice*, 11(10):1099–1111, 2003.

[118] H. L. Trentelman, A. A. Stoorvogel, and M. Hautus. *Control Theory for Linear Systems*. Springer-Verlag, London, 2001.

[119] S. Trimpe and R. D'Andrea. An experimental demonstration of a distributed and event-based state estimation algorithm. In *Proceedings of IFAC World Congress*, Milano, 2011 (to appear).

[120] P. Varutti, T. Faulwasser, B. Kern, M. Kögel, and R. Findeisen. Event-based reduced-attention predictive control for nonlinear uncertain systems. In *Proceedings of Computer-Aided Control System Design*, pages 1085–1090, Yokohama, 2010.

[121] V. Vasyutynskyy and K. Kabitzsch. Implementation of PID controller with send-on-delta sampling. In *Proceedings of International Control Conference*, Glasgow, 2006.

[122] M. Velasco, P. Marti, and J. Fuertes. The self triggered task model for real-time control systems. In *Work-in-Progress Session of IEEE Real-Time Systems Symposium*, Cancun, 2003.

[123] F.-Y. Wang and D. Liu. *Networked Control Systems: Theory and Applications*. Springer-Verlag, London, 2008.

[124] X. Wang and M. D. Lemmon. Event design in event-triggered feedback control systems. In *Proceedings of IEEE Conference on Decision and Control*, pages 2105–2110, Cancun, 2008.

[125] X. Wang and M. D. Lemmon. Event-triggered broadcasting across distributed networked control systems. In *Proceedings of American Control Conference*, pages 3139–3144, Seattle, 2008.

[126] X. Wang and M. D. Lemmon. Self-triggered feedback control systems with finite-gain L_2 stability. *IEEE Transactions on Automatic Control*, 54(3):452–467, 2009.

[127] X. Wang and M. D. Lemmon. Event-triggering in distributed networked systems with data dropouts and delays. In *Hybrid Systems: Computation and Control*, pages 366–380, San Francisco, 2009.

[128] X. Wang and M. D. Lemmon. Self-triggering under state-independent disturbances. *IEEE Transactions on Automatic Control*, 55(6):1494–1500, 2010.

[129] X. Wang and M. D. Lemmon. Event-triggering in distributed networked control systems. *IEEE Transactions on Automatic Control*, 53(3):586–601, 2011.

[130] W. S. Wong and R. W. Brockett. Systems with finite communication bandwidth constraints: stabilization with limited information feedback. *IEEE Transactions on Automatic Control*, 44(5):1049–1052, 1999.

[131] T. C. Yang. Networked control systems: a brief survey. *IEE Proceedings: Control Theory and Applications*, 153(4):403–412, 2006.

[132] W. Zhang, M. S. Branicky, and S. M. Philips. Stability of networked control systems. *IEEE Control Systems Magazine*, 21(1):84–99, 2001.

Appendix

A. Nonlinear model of the thermofluid process

The nonlinear model of the thermofluid process depicted in Fig. 2.3 is given by:

$$\begin{pmatrix} \dot{x}_1(t) \\ \dot{x}_2(t) \end{pmatrix} = \begin{pmatrix} \frac{1}{A}\big(q_{\mathrm{T3}}(u_1(t)) + q_{\mathrm{HW}}(d(t)) - q_{\mathrm{A}}(x_1(t), u_4(t))\big) \\ \frac{1}{A \cdot x_1(t)}\Big(q_{\mathrm{T3}}(u_1(t))(\vartheta_{\mathrm{T3}} - x_2(t)) + q_{\mathrm{HW}}(d(t))(\vartheta_{\mathrm{HW}} - x_2(t)) + \ldots \\ \qquad \ldots + \frac{k_{\mathrm{h}} P_{\mathrm{el}} u_2(t)}{\varrho\, c_{\mathrm{p}}}\Big) \end{pmatrix}$$

$$q_{\mathrm{T3}}(u_1(t)) = \begin{cases} 7 \times 10^{-6} \cdot (11.1 \cdot (u_1(t))^2 + 13.1 \cdot u_1(t) + 0.2) & , 0.05 \le u_1 \le 1; \\ 0 & , 0 \le u_1 < 0.05; \end{cases}$$

$$q_{\mathrm{HW}}(d(t)) = \begin{cases} 9.8 \times 10^{-5} & , 0.27 \le d \le 1; \\ 7 \times 10^{-4} \cdot (-2.6 \cdot (d(t))^2 + 1.4 \cdot d(t) - 0.05) & , 0.05 \le d < 0.27; \\ 0 & , 0 \le d < 0.05; \end{cases}$$

$$q_{\mathrm{A}}(x_1(t), u_4(t)) = K_{\mathrm{TB}}(u_4(t))\sqrt{2 g x_1(t)}$$

$$K_{\mathrm{TB}}(u_4(t)) = \begin{cases} 7 \times 10^{-6} \cdot (1.7 \cdot (u_4(t))^2 + 13 \cdot u_4(t) - 0.4) & , 0.1 \le u_4 \le 1; \\ 0 & , 0 \le u_4 < 0.1. \end{cases}$$

The characters q_i denote specific volume flows which have the unit m^3/s. Here, q_{T3} denotes the flow from the water supply T3 into tank TB through the valve V_1 and q_{A} denotes the outflow of TB over valve V_3 with the specific valve parameter K_{TB} (m^3/m). The flow q_{HW} denotes the inflow from the water supply HW. Moreover, ϑ_{T3} and ϑ_{HW} denote the fluid temperatures provided by the water supplies T3 and HW both of which are constant. The parameters are listed in Tab. A.1.

Table A.1.: Parameters

Parameter	Value	Meaning
P_{el}	3000 W	Heating power
A	$0.07\ \text{m}^2$	Cross sectional area of tank TB
k_{h}	$0.7\ \frac{\text{J}}{\text{W s}}$	Heating coefficient
c_{p}	$4180\ \frac{\text{J}}{\text{kg K}}$	Heat capacity of water
g	$9.81\ \frac{\text{m}}{\text{s}^2}$	Gravitation constant
ϱ	$998\ \frac{\text{kg}}{\text{m}^3}$	Density of water
ϑ_{HW}	$294.15\ \text{K}$	Fluid temperature of the water supply HW
ϑ_{T3}	$294.15\ \text{K}$	Fluid temperature of the water supply T3

B. Proofs

B.1. Proof of Lemma 3

In the time interval $[0, t_1)$ before the first event, the event-based control loop (3.1), (3.2), (3.12), (3.13), (3.21), (3.23), (3.24) subject to a constant disturbance $\bar{\boldsymbol{d}}$ is described by

$$\begin{aligned}\begin{pmatrix}\dot{\boldsymbol{x}}(t)\\ \dot{\boldsymbol{x}}_{\mathrm{s}}(t)\end{pmatrix} &= \begin{pmatrix}\boldsymbol{A} & -\boldsymbol{BK}\\ \boldsymbol{O} & \bar{\boldsymbol{A}}\end{pmatrix}\begin{pmatrix}\boldsymbol{x}(t)\\ \boldsymbol{x}_{\mathrm{s}}(t)\end{pmatrix} + \begin{pmatrix}\boldsymbol{E}\\ \boldsymbol{O}\end{pmatrix}\bar{\boldsymbol{d}} + \begin{pmatrix}\boldsymbol{BV}\\ \boldsymbol{BV}\end{pmatrix}\boldsymbol{w}(t)\\ \begin{pmatrix}\boldsymbol{x}(0)\\ \boldsymbol{x}_{\mathrm{s}}(0)\end{pmatrix} &= \begin{pmatrix}\boldsymbol{x}_0\\ \boldsymbol{x}_0\end{pmatrix}.\end{aligned}$$

Using state transformation (3.14), the following model is obtained

$$\begin{pmatrix}\dot{\boldsymbol{x}}_{\Delta}(t)\\ \dot{\boldsymbol{x}}_{\mathrm{s}}(t)\end{pmatrix} = \begin{pmatrix}\boldsymbol{A} & \boldsymbol{O}\\ \boldsymbol{O} & \bar{\boldsymbol{A}}\end{pmatrix}\begin{pmatrix}\boldsymbol{x}_{\Delta}(t)\\ \boldsymbol{x}_{\mathrm{s}}(t)\end{pmatrix} + \begin{pmatrix}\boldsymbol{E}\\ \boldsymbol{O}\end{pmatrix}\bar{\boldsymbol{d}} + \begin{pmatrix}\boldsymbol{O}\\ \boldsymbol{BV}\end{pmatrix}\bar{\boldsymbol{w}} \tag{B.1}$$

$$\begin{pmatrix}\boldsymbol{x}_{\Delta}(0)\\ \boldsymbol{x}_{\mathrm{s}}(0)\end{pmatrix} = \begin{pmatrix}0\\ \boldsymbol{x}_0\end{pmatrix}. \tag{B.2}$$

When at time t_1 equality (3.21) is met, the state $\boldsymbol{x}_{\Delta}(t_1)$ is given by

$$\boldsymbol{x}_{\Delta}(t_1) = \int_0^{t_1} \mathrm{e}^{\boldsymbol{A}(t-\alpha)}\boldsymbol{E}\bar{\boldsymbol{d}}\,\mathrm{d}\alpha = \boldsymbol{A}^{-1}\left(\mathrm{e}^{\boldsymbol{A}t_1} - \boldsymbol{I}_n\right)\boldsymbol{E}\bar{\boldsymbol{d}}$$

and, hence, the disturbance estimate (3.24) satisfies

$$\begin{aligned}\hat{\boldsymbol{d}}_1 &= \hat{\boldsymbol{d}}_0 + \left(\boldsymbol{A}^{-1}\left(\mathrm{e}^{\boldsymbol{A}(t_1-t_0)} - \boldsymbol{I}_n\right)\boldsymbol{E}\right)^{+}\boldsymbol{x}_{\Delta}(t_1)\\ &= \left(\boldsymbol{A}^{-1}\left(\mathrm{e}^{\boldsymbol{A}t_1} - \boldsymbol{I}_n\right)\boldsymbol{E}\right)^{+}\cdot\left(\boldsymbol{A}^{-1}\left(\mathrm{e}^{\boldsymbol{A}t_1} - \boldsymbol{I}_n\right)\boldsymbol{E}\right)\bar{\boldsymbol{d}}\\ &= \bar{\boldsymbol{d}},\end{aligned}$$

which proves the lemma. □

B.2. Proof of Lemma 5

According to Eq. (3.21) there will be no event generated as long as the inequality

$$\|\boldsymbol{x}(t) - \boldsymbol{x}_s(t)\| < \bar{e} \tag{B.3}$$

holds for all $t > 0$. According to Eq. (3.20), which is applied for $k = 0$ and $\hat{\boldsymbol{d}}_0 = \boldsymbol{0}$, the inequality (B.3) is true if

$$\left\| \int_0^t \mathrm{e}^{\boldsymbol{A}(t-\alpha)} \boldsymbol{E} \left(\bar{d}\tilde{\boldsymbol{d}}(\alpha) - \hat{\boldsymbol{d}}_0\right) \mathrm{d}\alpha \right\| < \bar{e}$$

holds for all times $t \geq 0$. An upper bound on the left-hand side of this inequality exists if the plant is stable, because then

$$\begin{aligned} \max_{t\geq 0} \left\| \int_0^t \mathrm{e}^{\boldsymbol{A}(t-\alpha)} \boldsymbol{E}\, \bar{d}\tilde{\boldsymbol{d}}(\alpha)\, \mathrm{d}\alpha \right\| &\leq \max_{t\geq 0} \int_0^t \left\| \mathrm{e}^{\boldsymbol{A}(t-\alpha)} \boldsymbol{E} \right\| \mathrm{d}\alpha \cdot \max_{t\geq 0} |\bar{d}\tilde{\boldsymbol{d}}(t)| \\ &\leq \int_0^\infty \left\| \mathrm{e}^{\boldsymbol{A}\alpha} \boldsymbol{E} \right\| \mathrm{d}\alpha \cdot |\bar{d}| \end{aligned}$$

leads to the inequality

$$|\bar{d}| < \frac{\bar{e}}{\int_0^\infty \|\mathrm{e}^{\boldsymbol{A}\alpha}\boldsymbol{E}\|\, \mathrm{d}\alpha} = \bar{d}_{\mathrm{UD}} \neq 0.$$

Thus, if the disturbance parameter $\bar{d}$ is chosen such that this inequality is satisfied, no event is generated, which proves the lemma. □

B.3. Proof of Lemma 6

According to Lemma 5, no event is generated in the undisturbed closed-loop system. Hence, Eqs. (B.1), (B.2) hold for all times $t \geq 0$ with $\bar{\boldsymbol{d}} = \boldsymbol{0}$. Due to the selection (3.7) of the matrix $\boldsymbol{V}$ for the output

$$\boldsymbol{y}(t) = \boldsymbol{y}_\Delta(t) + \boldsymbol{y}_s(t)$$

with

$$\begin{aligned} \boldsymbol{y}_\Delta(t) &= \boldsymbol{C}\boldsymbol{x}_\Delta(t) \\ \boldsymbol{y}_\mathrm{s}(t) &= \boldsymbol{C}\boldsymbol{x}_\mathrm{s}(t), \end{aligned}$$

the relation

$$\lim_{t\to\infty} \|\boldsymbol{y}_\mathrm{s}(t) - \bar{\boldsymbol{w}}\| = 0 \tag{B.4}$$

is obtained. This can be seen from the second line of Eq. (B.1). As the initial condition of the difference state $\boldsymbol{x}_\Delta$ is zero ($\boldsymbol{x}_\Delta(0) = \boldsymbol{0}$) and $\boldsymbol{x}_\Delta(t)$ is not affected by any exogenous signal, $\boldsymbol{x}_\Delta(t)$ is zero for all times $t \geq 0$ (see first line of Eq. (B.1)) and, therefore,

$$\lim_{t\to\infty} \|\boldsymbol{y}(t) - \bar{\boldsymbol{w}}\| = 0$$

holds, which proves the lemma. □

B.4. Proof of Theorem 6

Due to Lemma 3 and Lemma 5, the event-based closed-loop system generates at most one event for constant exogenous signals (3.43).

If the plant is stable and the disturbance magnitude $\bar{\boldsymbol{d}}$ is small enough, no event is generated and Eqs. (B.1), (B.2) hold for all $t \geq 0$. Due to Eq. (B.4), the relation

$$\lim_{t\to\infty} \|\boldsymbol{y}(t) - \bar{\boldsymbol{w}}\| = \lim_{t\to\infty} \|\boldsymbol{y}_\Delta(t)\| = \|\boldsymbol{C}\boldsymbol{A}^{-1}\boldsymbol{E}\bar{\boldsymbol{d}}\|$$

results.

If an event is generated at time t_1, which always happens for an unstable plant affected by an arbitrarily small but nonzero disturbance (see first line of Eq. (B.1)), Eqs. (3.1), (3.13), (3.24) yield the following model for $t \geq t_1$

$$\begin{aligned} \begin{pmatrix} \dot{\boldsymbol{x}}(t) \\ \dot{\boldsymbol{x}}_\mathrm{s}(t) \end{pmatrix} &= \begin{pmatrix} \boldsymbol{A} & -\boldsymbol{B}\boldsymbol{K} \\ \boldsymbol{O} & \bar{\boldsymbol{A}} \end{pmatrix} \begin{pmatrix} \boldsymbol{x}(t) \\ \boldsymbol{x}_\mathrm{s}(t) \end{pmatrix} + \begin{pmatrix} \boldsymbol{E} \\ \boldsymbol{E} \end{pmatrix} \bar{\boldsymbol{d}} + \begin{pmatrix} \boldsymbol{B}\boldsymbol{V} \\ \boldsymbol{B}\boldsymbol{V} \end{pmatrix} \bar{\boldsymbol{w}} \\ \begin{pmatrix} \boldsymbol{x}(t_1) \\ \boldsymbol{x}_\mathrm{s}(t_1^+) \end{pmatrix} &= \begin{pmatrix} \boldsymbol{x}(t_1) \\ \boldsymbol{x}(t_1) \end{pmatrix}, \end{aligned}$$

which after state transformation (3.14) results in

$$\begin{pmatrix} \dot{\boldsymbol{x}}_\Delta(t) \\ \dot{\boldsymbol{x}}_{\mathrm{s}}(t) \end{pmatrix} = \begin{pmatrix} \boldsymbol{A} & \boldsymbol{O} \\ \boldsymbol{O} & \bar{\boldsymbol{A}} \end{pmatrix} \begin{pmatrix} \boldsymbol{x}_\Delta(t) \\ \boldsymbol{x}_{\mathrm{s}}(t) \end{pmatrix} + \begin{pmatrix} \boldsymbol{O} \\ \boldsymbol{E} \end{pmatrix} \bar{\boldsymbol{d}} + \begin{pmatrix} \boldsymbol{O} \\ \boldsymbol{BV} \end{pmatrix} \bar{\boldsymbol{w}} \tag{B.5}$$

$$\begin{pmatrix} \boldsymbol{x}_\Delta(t_1^+) \\ \boldsymbol{x}_{\mathrm{s}}(t_1^+) \end{pmatrix} = \begin{pmatrix} \boldsymbol{0} \\ \boldsymbol{x}(t_1) \end{pmatrix}.$$

Due to the choice of the matrix $\boldsymbol{V}$ and $\boldsymbol{x}_\Delta(t)$ being zero for all times $t \geq t_1$ (see first line in Eq. (B.5)), the relations

$$\lim_{t\to\infty} \boldsymbol{y}_{\mathrm{s}}(t) = \bar{\boldsymbol{w}} - \boldsymbol{C}\bar{\boldsymbol{A}}^{-1}\boldsymbol{E}\bar{\boldsymbol{d}}$$

and

$$\lim_{t\to\infty} \|\boldsymbol{y}(t) - \bar{\boldsymbol{w}}\| = \lim_{t\to\infty} \|\boldsymbol{y}_\Delta(t) + \boldsymbol{y}_{\mathrm{s}}(t) - \bar{\boldsymbol{w}}\| = \|\boldsymbol{C}\bar{\boldsymbol{A}}^{-1}\boldsymbol{E}\bar{\boldsymbol{d}}\|$$

hold. The summary of both cases proves the theorem. □

B.5. Proof of Theorem 8

The difference $\boldsymbol{e}_{\mathrm{DT}}(t) = \boldsymbol{x}_{\mathrm{DT}}(t) - \boldsymbol{x}_{\mathrm{CT}}(t)$ between the state $\boldsymbol{x}_{\mathrm{DT}}(t)$ of the discrete-time state-feedback loop (3.56), (3.57) and the state $\boldsymbol{x}_{\mathrm{CT}}(t)$ of the continuous-time control loop (3.5), (3.6) is given by

$$\dot{\boldsymbol{e}}_{\mathrm{DT}}(t) = \bar{\boldsymbol{A}}\boldsymbol{e}_{\mathrm{DT}}(t) + \boldsymbol{BK}\boldsymbol{x}_{\Delta,\mathrm{DT}}(t), \quad \boldsymbol{e}_{\mathrm{DT}}(0) = \boldsymbol{0}.$$

As $\bar{\boldsymbol{A}}$ is assumed to be Hurwitz, the error $\boldsymbol{e}_{\mathrm{DT}}(t)$ is bounded if $\boldsymbol{x}_{\Delta,\mathrm{DT}}(t)$ is bounded. According to relations (3.54)–(3.56), $\dot{\boldsymbol{x}}_{\Delta,\mathrm{DT}}(t)$ is given for $t \in [\ell T_{\mathrm{s}}, (\ell+1)T_{\mathrm{s}})$ by

$$\begin{aligned} \dot{\boldsymbol{x}}_{\Delta,\mathrm{DT}}(t) &= \dot{\boldsymbol{x}}_{\mathrm{DT}}(t) - \dot{\boldsymbol{x}}_{\mathrm{s,DT}}(t) \\ &= \boldsymbol{A}\boldsymbol{x}_{\mathrm{DT}}(t) - \boldsymbol{BK}\boldsymbol{x}_{\mathrm{s,DT}}(t) + \boldsymbol{E}\boldsymbol{d}(t) \\ &= \boldsymbol{A}\boldsymbol{x}_{\Delta,\mathrm{DT}}(t) + (\boldsymbol{A} - \boldsymbol{BK})\boldsymbol{x}(\ell T_{\mathrm{s}}) + \boldsymbol{E}\boldsymbol{d}(t), \quad \boldsymbol{x}_{\Delta,\mathrm{DT}}(\ell T_{\mathrm{s}}) = \boldsymbol{0}. \end{aligned}$$

Hence, with

$$x_{\mathrm{DT,max}} = \max_{\ell\in\{0,1,\ldots,\infty\}} \|\boldsymbol{x}(\ell T_{\mathrm{s}})\|,$$

the difference state $\boldsymbol{x}_{\Delta,\mathrm{DT}}(t)$ is bounded by

$$\begin{aligned}\|\boldsymbol{x}_{\Delta,\mathrm{DT}}(t)\| &= \left\| \int_0^{T_s} \mathrm{e}^{\boldsymbol{A}(t-\alpha)} (\bar{\boldsymbol{A}}\boldsymbol{x}(\ell T_s) + \boldsymbol{E}\boldsymbol{d}(\alpha))\,\mathrm{d}\alpha \right\| \\ &\leq \int_0^{T_s} \left\| \mathrm{e}^{\boldsymbol{A}\alpha} \right\| \mathrm{d}\alpha \cdot (\|\bar{\boldsymbol{A}}\| x_{\mathrm{DT,max}} + \|\boldsymbol{E}\| d_{\max}) = x_{\Delta\max}\end{aligned}$$

if $x_{\mathrm{DT,max}}$ is bounded. To derive the state bound $x_{\mathrm{DT,max}}$, the discrete-time model

$$\boldsymbol{x}(\ell+1) = \boldsymbol{A}_{\mathrm{DT}}\boldsymbol{x}(\ell) + \boldsymbol{B}_{\mathrm{DT}}\boldsymbol{u}(\ell) + \boldsymbol{E}_{\mathrm{DT}}\boldsymbol{d}(\ell), \qquad \boldsymbol{x}(0) = \boldsymbol{x}_0 \tag{B.6}$$

$$\boldsymbol{y}(\ell) = \boldsymbol{C}_{\mathrm{DT}}\boldsymbol{x}(\ell) \tag{B.7}$$

of plant (3.1), (3.2) with

$$\begin{aligned}\boldsymbol{A}_{\mathrm{DT}} &= \mathrm{e}^{\boldsymbol{A}T_s} \\ \boldsymbol{B}_{\mathrm{DT}} &= \int_0^{T_s} \mathrm{e}^{\boldsymbol{A}\alpha}\boldsymbol{B}\,\mathrm{d}\alpha \\ \boldsymbol{E}_{\mathrm{DT}} &= \int_0^{T_s} \mathrm{e}^{\boldsymbol{A}\alpha}\boldsymbol{E}\,\mathrm{d}\alpha \\ \boldsymbol{C}_{\mathrm{DT}} &= \boldsymbol{C}\end{aligned}$$

is considered [83]. With plant (B.6), (B.7) and the controller $\boldsymbol{u}(\ell) = -\boldsymbol{K}\boldsymbol{x}(\ell)$ the discrete-time state-feedback loop (3.56), (3.57) can be alternatively described by the discrete-time model

$$\boldsymbol{x}_{\mathrm{DT}}(\ell+1) = \underbrace{(\boldsymbol{A}_{\mathrm{DT}} - \boldsymbol{B}_{\mathrm{DT}}\boldsymbol{K})}_{\bar{\boldsymbol{A}}_{\mathrm{DT}}}\boldsymbol{x}_{\mathrm{DT}}(\ell) + \boldsymbol{E}\boldsymbol{d}(\ell), \quad \boldsymbol{x}_{\mathrm{DT}}(0) = \boldsymbol{x}_0 \tag{B.8}$$

$$\boldsymbol{y}_{\mathrm{DT}}(\ell) = \boldsymbol{C}_{\mathrm{DT}}\boldsymbol{x}_{\mathrm{DT}}(\ell), \tag{B.9}$$

where in the following the disturbance is assumed to be constant, i.e. $\boldsymbol{d}(\ell) = \bar{\boldsymbol{d}}$.

Theorem 19. [32, 83] *The state $\boldsymbol{x}_{\mathrm{DT}}(t)$ of the discrete-time state-feedback loop* (B.8), (B.9) *is GUUB if*

$$\left|\lambda_i\left\{\bar{\boldsymbol{A}}_{\mathrm{DT}}\right\}\right| < 1$$

holds for all $i \in \{1, 2, ..., n\}$ and the disturbance $\boldsymbol{d}(\ell)$ is bounded.

Consequently, if $\|\bar{\boldsymbol{A}}_{\mathrm{DT}}\| < 1$ holds, $x_{\mathrm{DT,max}}$ exists and can be obtained according to

$$\begin{aligned}
\|\boldsymbol{x}_{\mathrm{DT}}(\ell)\| &= \left\| \bar{\boldsymbol{A}}_{\mathrm{DT}}^{\ell}\boldsymbol{x}_0 + \sum_{j=0}^{\ell-1} \bar{\boldsymbol{A}}_{\mathrm{DT}}^{\ell-1-j}\boldsymbol{E}_{\mathrm{DT}}\boldsymbol{d}(j) \right\| \\
&\leq \max_{\ell\in\{0,1,\dots,\infty\}} \left\|\bar{\boldsymbol{A}}_{\mathrm{DT}}^{\ell}\right\| \cdot \|\boldsymbol{x}_0\| + \sum_{j=0}^{\infty} \|\bar{\boldsymbol{A}}_{\mathrm{DT}}\|^{j} \|\boldsymbol{E}_{\mathrm{DT}}\| d_{\max} \\
&= \max_{\ell\in\{0,1,\dots,\infty\}} \left\|\bar{\boldsymbol{A}}_{\mathrm{DT}}^{\ell}\right\| \cdot \|\boldsymbol{x}_0\| + \frac{\|\boldsymbol{E}_{\mathrm{DT}}\|\, d_{\max}}{1-\|\bar{\boldsymbol{A}}_{\mathrm{DT}}\|} = x_{\mathrm{DT,max}},
\end{aligned}$$

which proves the theorem. □

B.6. Proof of Theorem 12

First, it is shown that Lemma 3 also holds for the event-based PI-control loop. To do so, the model of the closed-loop system (4.18), (4.20), (4.21), (4.24), (4.25) is written down for the time interval $t \leq t_1$:

$$\begin{aligned}
\begin{pmatrix} \dot{\boldsymbol{x}}(t) \\ \dot{\boldsymbol{x}}_{\mathrm{s}}(t) \\ \dot{\boldsymbol{x}}_{\mathrm{sr}}(t) \end{pmatrix} &= \begin{pmatrix} \boldsymbol{A} & -\boldsymbol{B}\boldsymbol{K}_{\mathrm{P}} & -\boldsymbol{B}\boldsymbol{K}_{\mathrm{I}} \\ \boldsymbol{O} & \boldsymbol{A}-\boldsymbol{B}\boldsymbol{K}_{\mathrm{P}} & -\boldsymbol{B}\boldsymbol{K}_{\mathrm{I}} \\ \boldsymbol{O} & \boldsymbol{C} & \boldsymbol{O} \end{pmatrix} \begin{pmatrix} \boldsymbol{x}(t) \\ \boldsymbol{x}_{\mathrm{s}}(t) \\ \boldsymbol{x}_{\mathrm{sr}}(t) \end{pmatrix} + \begin{pmatrix} \boldsymbol{O} \\ \boldsymbol{O} \\ -\boldsymbol{I}_r \end{pmatrix} \bar{\boldsymbol{w}} + \begin{pmatrix} \boldsymbol{E} \\ \boldsymbol{O} \\ \boldsymbol{O} \end{pmatrix} \bar{\boldsymbol{d}} \\
\begin{pmatrix} \boldsymbol{x}(0) \\ \boldsymbol{x}_{\mathrm{s}}(0) \\ \boldsymbol{x}_{\mathrm{sr}}(0) \end{pmatrix} &= \begin{pmatrix} \boldsymbol{x}_0 \\ \boldsymbol{x}_0 \\ \boldsymbol{x}_{\mathrm{sr}0} \end{pmatrix} \\
\boldsymbol{y}(t) &= \begin{pmatrix} \boldsymbol{C} & \boldsymbol{O} & \boldsymbol{O} \end{pmatrix} \begin{pmatrix} \boldsymbol{x}(t) \\ \boldsymbol{x}_{\mathrm{s}}(t) \\ \boldsymbol{x}_{\mathrm{sr}}(t) \end{pmatrix}.
\end{aligned}$$

With the state transformation

$$\begin{pmatrix} \boldsymbol{x}_{\Delta}(t) \\ \boldsymbol{x}_{\mathrm{s}}(t) \\ \boldsymbol{x}_{\mathrm{sr}}(t) \end{pmatrix} = \begin{pmatrix} \boldsymbol{I}_n & -\boldsymbol{I}_n & \boldsymbol{O} \\ \boldsymbol{O} & \boldsymbol{I}_n & \boldsymbol{O} \\ \boldsymbol{O} & \boldsymbol{O} & \boldsymbol{I}_r \end{pmatrix} \begin{pmatrix} \boldsymbol{x}(t) \\ \boldsymbol{x}_{\mathrm{s}}(t) \\ \boldsymbol{x}_{\mathrm{sr}}(t) \end{pmatrix} \tag{B.10}$$

the model

$$\begin{pmatrix} \dot{\boldsymbol{x}}_\Delta(t) \\ \hline \dot{\boldsymbol{x}}_{\mathrm{s}}(t) \\ \dot{\boldsymbol{x}}_{\mathrm{sr}}(t) \end{pmatrix} = \left(\begin{array}{c|cc} \boldsymbol{A} & \boldsymbol{O} & \boldsymbol{O} \\ \hline \boldsymbol{O} & \boldsymbol{A}-\boldsymbol{B}\boldsymbol{K}_{\mathrm{P}} & -\boldsymbol{B}\boldsymbol{K}_{\mathrm{I}} \\ \boldsymbol{O} & \boldsymbol{C} & \boldsymbol{O} \end{array}\right) \begin{pmatrix} \boldsymbol{x}_\Delta(t) \\ \hline \boldsymbol{x}_{\mathrm{s}}(t) \\ \boldsymbol{x}_{\mathrm{sr}}(t) \end{pmatrix} + \begin{pmatrix} \boldsymbol{O} \\ \hline \boldsymbol{O} \\ -\boldsymbol{I}_r \end{pmatrix} \bar{\boldsymbol{w}} + \begin{pmatrix} \boldsymbol{E} \\ \hline \boldsymbol{O} \\ \boldsymbol{O} \end{pmatrix} \bar{\boldsymbol{d}}$$

$$\begin{pmatrix} \boldsymbol{x}_\Delta(0) \\ \hline \boldsymbol{x}_{\mathrm{s}}(0) \\ \boldsymbol{x}_{\mathrm{sr}}(0) \end{pmatrix} = \begin{pmatrix} \boldsymbol{0} \\ \hline \boldsymbol{x}_0 \\ \boldsymbol{x}_{\mathrm{sr}0} \end{pmatrix}$$

$$\boldsymbol{y}(t) = \left(\begin{array}{c|cc} \boldsymbol{C} & \boldsymbol{C} & \boldsymbol{O} \end{array}\right) \begin{pmatrix} \boldsymbol{x}_\Delta(t) \\ \hline \boldsymbol{x}_{\mathrm{s}}(t) \\ \boldsymbol{x}_{\mathrm{sr}}(t) \end{pmatrix}$$

is obtained, which can be rewritten in the compact form

$$\begin{pmatrix} \dot{\boldsymbol{x}}_\Delta(t) \\ \dot{\boldsymbol{x}}_{\mathrm{sI}}(t) \end{pmatrix} = \begin{pmatrix} \boldsymbol{A} & \boldsymbol{O} \\ \boldsymbol{O} & \bar{\boldsymbol{A}}_{\mathrm{I}} \end{pmatrix} \begin{pmatrix} \boldsymbol{x}_\Delta(t) \\ \boldsymbol{x}_{\mathrm{sI}}(t) \end{pmatrix} + \begin{pmatrix} \boldsymbol{O} \\ \boldsymbol{F}_{\mathrm{I}} \end{pmatrix} \bar{\boldsymbol{w}} + \begin{pmatrix} \boldsymbol{E} \\ \boldsymbol{O} \end{pmatrix} \bar{\boldsymbol{d}}$$

$$\begin{pmatrix} \boldsymbol{x}_\Delta(0) \\ \boldsymbol{x}_{\mathrm{sI}}(0) \end{pmatrix} = \begin{pmatrix} \boldsymbol{0} \\ \begin{pmatrix} \boldsymbol{x}_0 \\ \boldsymbol{x}_{\mathrm{sr}0} \end{pmatrix} \end{pmatrix}$$

$$\boldsymbol{y}(t) = \begin{pmatrix} \boldsymbol{C} & \boldsymbol{C}_{\mathrm{I}} \end{pmatrix} \begin{pmatrix} \boldsymbol{x}_\Delta(t) \\ \boldsymbol{x}_{\mathrm{sI}}(t) \end{pmatrix}.$$

The matrices $\bar{\boldsymbol{A}}_{\mathrm{I}}$, $\boldsymbol{C}_{\mathrm{I}}$ and $\boldsymbol{F}_{\mathrm{I}}$ are defined in Eqs. (4.22), (4.23). The first line of this model coincides with the first line of Eq. (B.1) used in the proof of Lemma 3. Hence, the equality $\hat{\boldsymbol{d}}_1 = \bar{\boldsymbol{d}}$ is also valid for the event-based PI-control loop.

Now, the behaviour of the closed-loop system after the first event is investigated. Equations (3.26), (4.18), (4.20), (4.21), (4.24), (4.25) yield for $t \geq t_1$

$$\begin{pmatrix} \dot{\boldsymbol{x}}(t) \\ \dot{\boldsymbol{x}}_{\mathrm{s}}(t) \\ \dot{\boldsymbol{x}}_{\mathrm{sr}}(t) \end{pmatrix} = \begin{pmatrix} \boldsymbol{A} & -\boldsymbol{B}\boldsymbol{K}_{\mathrm{P}} & -\boldsymbol{B}\boldsymbol{K}_{\mathrm{I}} \\ \boldsymbol{O} & \boldsymbol{A}-\boldsymbol{B}\boldsymbol{K}_{\mathrm{P}} & -\boldsymbol{B}\boldsymbol{K}_{\mathrm{I}} \\ \boldsymbol{O} & \boldsymbol{C} & \boldsymbol{O} \end{pmatrix} \begin{pmatrix} \boldsymbol{x}(t) \\ \boldsymbol{x}_{\mathrm{s}}(t) \\ \boldsymbol{x}_{\mathrm{sr}}(t) \end{pmatrix} + \begin{pmatrix} \boldsymbol{O} \\ \boldsymbol{O} \\ -\boldsymbol{I}_r \end{pmatrix} \bar{\boldsymbol{w}} + \begin{pmatrix} \boldsymbol{E} \\ \boldsymbol{E} \\ \boldsymbol{O} \end{pmatrix} \bar{\boldsymbol{d}}$$

$$\begin{pmatrix} \boldsymbol{x}(t_1) \\ \boldsymbol{x}_{\mathrm{s}}(t_1^+) \\ \boldsymbol{x}_{\mathrm{sr}}(t_1) \end{pmatrix} = \begin{pmatrix} \boldsymbol{x}(t_1) \\ \boldsymbol{x}(t_1) \\ \boldsymbol{x}_{\mathrm{sr}}(t_1) \end{pmatrix}$$

with the output equation

$$\boldsymbol{y}(t) = \begin{pmatrix} \boldsymbol{C} & \boldsymbol{O} & \boldsymbol{O} \end{pmatrix} \begin{pmatrix} \boldsymbol{x}(t) \\ \boldsymbol{x}_{\mathrm{s}}(t) \\ \boldsymbol{x}_{\mathrm{sr}}(t) \end{pmatrix}.$$

Applying state transformation (B.10), the model

$$\begin{aligned}
\begin{pmatrix} \dot{\boldsymbol{x}}_{\Delta}(t) \\ \hline \dot{\boldsymbol{x}}_{\mathrm{s}}(t) \\ \dot{\boldsymbol{x}}_{\mathrm{sr}}(t) \end{pmatrix} &= \left(\begin{array}{c|cc} \boldsymbol{A} & \boldsymbol{O} & \boldsymbol{O} \\ \hline \boldsymbol{O} & \boldsymbol{A}-\boldsymbol{B}\boldsymbol{K}_{\mathrm{P}} & -\boldsymbol{B}\boldsymbol{K}_{\mathrm{I}} \\ \boldsymbol{O} & \boldsymbol{C} & \boldsymbol{O} \end{array}\right) \begin{pmatrix} \boldsymbol{x}_{\Delta}(t) \\ \hline \boldsymbol{x}_{\mathrm{s}}(t) \\ \boldsymbol{x}_{\mathrm{sr}}(t) \end{pmatrix} + \begin{pmatrix} \boldsymbol{O} \\ \hline \boldsymbol{O} \\ -\boldsymbol{I}_r \end{pmatrix} \bar{\boldsymbol{w}} + \begin{pmatrix} \boldsymbol{O} \\ \hline \boldsymbol{E} \\ \boldsymbol{O} \end{pmatrix} \bar{\boldsymbol{d}} \\
\begin{pmatrix} \boldsymbol{x}_{\Delta}(t_1^+) \\ \hline \boldsymbol{x}_{\mathrm{s}}(t_1^+) \\ \boldsymbol{x}_{\mathrm{sr}}(t_1) \end{pmatrix} &= \begin{pmatrix} \boldsymbol{0} \\ \hline \boldsymbol{x}(t_1) \\ \boldsymbol{x}_{\mathrm{sr}}(t_1) \end{pmatrix} \\
\boldsymbol{y}(t) &= \left(\begin{array}{c|cc} \boldsymbol{C} & \boldsymbol{C} & \boldsymbol{O} \end{array}\right) \begin{pmatrix} \boldsymbol{x}_{\Delta}(t) \\ \hline \boldsymbol{x}_{\mathrm{s}}(t) \\ \boldsymbol{x}_{\mathrm{sr}}(t) \end{pmatrix}
\end{aligned}$$

and, therefore,

$$\begin{aligned}
\begin{pmatrix} \dot{\boldsymbol{x}}_{\Delta}(t) \\ \dot{\boldsymbol{x}}_{\mathrm{sI}}(t) \end{pmatrix} &= \begin{pmatrix} \boldsymbol{A} & \boldsymbol{O} \\ \boldsymbol{O} & \bar{\boldsymbol{A}}_{\mathrm{I}} \end{pmatrix} \begin{pmatrix} \boldsymbol{x}_{\Delta}(t) \\ \boldsymbol{x}_{\mathrm{sI}}(t) \end{pmatrix} + \begin{pmatrix} \boldsymbol{O} \\ \boldsymbol{F}_{\mathrm{I}} \end{pmatrix} \bar{\boldsymbol{w}} + \begin{pmatrix} \boldsymbol{O} \\ \boldsymbol{E}_{\mathrm{I}} \end{pmatrix} \bar{\boldsymbol{d}} \\
\begin{pmatrix} \boldsymbol{x}_{\Delta}(t_1^+) \\ \boldsymbol{x}_{\mathrm{sI}}(t_1^+) \end{pmatrix} &= \begin{pmatrix} \boldsymbol{0} \\ \begin{pmatrix} \boldsymbol{x}(t_1) \\ \boldsymbol{x}_{\mathrm{sr}}(t_1) \end{pmatrix} \end{pmatrix} \\
\boldsymbol{y}(t) &= \begin{pmatrix} \boldsymbol{C} & \boldsymbol{C}_{\mathrm{I}} \end{pmatrix} \begin{pmatrix} \boldsymbol{x}_{\Delta}(t) \\ \boldsymbol{x}_{\mathrm{sI}}(t) \end{pmatrix}
\end{aligned}$$

results. In the last model neither the disturbance nor the reference signal has an influence on the state $\boldsymbol{x}_{\Delta}(t)$. Furthermore, the second part of the model coincides with the model (4.22), (4.23) of the continuous-time PI-control loop which has the setpoint tracking property (Theorem 11, page 78). Hence, the output $\boldsymbol{y}(t)$ asymptotically reaches the setpoint $\bar{\boldsymbol{w}}$. □

B.7. Proof of Lemma 9

For $\left\|\boldsymbol{d}(t) - \hat{\boldsymbol{d}}_k\right\| < \bar{d}_{\mathrm{UD}}$, no event is generated for all times $t \geq t_k$. Therefore, the bound

$$\begin{aligned} \|\boldsymbol{y}(t) - \boldsymbol{y}_{\mathrm{s}}(t)\| &= \|\boldsymbol{C}\boldsymbol{x}(t) - \boldsymbol{C}\boldsymbol{x}_{\mathrm{s}}(t)\| \\ &\leq \|\boldsymbol{C}\|\|\boldsymbol{x}(t) - \boldsymbol{x}_{\mathrm{s}}(t)\| < \|\boldsymbol{C}\|\bar{e} \end{aligned}$$

holds (Eq. (4.29)) and the relation $\boldsymbol{y}_{\mathrm{s}}(t) \to \bar{\boldsymbol{w}}$ is obtained for large t because $\hat{\boldsymbol{d}}_k$ is constant for $t \geq t_k$, which yields the set (4.31). □

B.8. Proof of Lemma 10

The difference $\boldsymbol{e}(t)$ is given by

$$\begin{aligned} \dot{\boldsymbol{e}}(t) &= \dot{\boldsymbol{x}}(t) - \dot{\boldsymbol{x}}_{\mathrm{CT}}(t) \\ &= \boldsymbol{A}\boldsymbol{x}(t) - \boldsymbol{B}\boldsymbol{K}_{\mathrm{P}}\boldsymbol{x}_{\mathrm{s}}(t) - \boldsymbol{B}\boldsymbol{K}_{\mathrm{I}}\boldsymbol{x}_{\mathrm{sr}}(t) + \boldsymbol{E}\boldsymbol{d}(t) \\ &\quad -\boldsymbol{A}\boldsymbol{x}_{\mathrm{CT}}(t) + \boldsymbol{B}\boldsymbol{K}_{\mathrm{P}}\boldsymbol{x}_{\mathrm{CT}}(t) + \boldsymbol{B}\boldsymbol{K}_{\mathrm{I}}\boldsymbol{x}_{\mathrm{CTr}}(t) - \boldsymbol{E}\boldsymbol{d}(t) \\ &= (\boldsymbol{A} - \boldsymbol{B}\boldsymbol{K}_{\mathrm{P}})\boldsymbol{e}(t) + \boldsymbol{B}\boldsymbol{K}_{\mathrm{P}}\boldsymbol{x}_{\Delta}(t) + \boldsymbol{B}\boldsymbol{K}_{\mathrm{I}}(\boldsymbol{x}_{\mathrm{CTr}}(t) - \boldsymbol{x}_{\mathrm{sr}}(t)). \end{aligned}$$

As $\boldsymbol{A} - \boldsymbol{B}\boldsymbol{K}_{\mathrm{P}}$ is assumed to be Hurwitz and $\boldsymbol{x}_{\Delta}(t)$ is bounded, the difference $\boldsymbol{e}(t)$ is bounded by

$$\|\boldsymbol{e}(t)\| \leq \int_0^{\infty} \left\|\mathrm{e}^{(\boldsymbol{A} - \boldsymbol{B}\boldsymbol{K}_{\mathrm{P}})\alpha}\right\| \mathrm{d}\alpha \cdot \left(\|\boldsymbol{B}\boldsymbol{K}_{\mathrm{P}}\|\bar{e} + \|\boldsymbol{B}\boldsymbol{K}_{\mathrm{I}}\|\|\boldsymbol{x}_{\mathrm{sr},\Delta}(t)\| \right) \tag{B.11}$$

if $\|\boldsymbol{x}_{\mathrm{sr},\Delta}(t)\| = \|\boldsymbol{x}_{\mathrm{CTr}}(t) - \boldsymbol{x}_{\mathrm{sr}}(t)\|$ is bounded. If Eqs. (4.22) and (4.24) are lumped together, they can be described for $t \geq t_k$ by

$$\begin{aligned} \begin{pmatrix} \dot{\boldsymbol{x}}_{\mathrm{CT}}(t) \\ \dot{\boldsymbol{x}}_{\mathrm{CTr}}(t) \\ \dot{\boldsymbol{x}}_{\mathrm{s}}(t) \\ \dot{\boldsymbol{x}}_{\mathrm{sr}}(t) \end{pmatrix} &= \begin{pmatrix} \boldsymbol{A} - \boldsymbol{B}\boldsymbol{K}_{\mathrm{P}} & -\boldsymbol{B}\boldsymbol{K}_{\mathrm{I}} & \boldsymbol{O} & \boldsymbol{O} \\ \boldsymbol{C} & \boldsymbol{O} & \boldsymbol{O} & \boldsymbol{O} \\ \boldsymbol{O} & \boldsymbol{O} & \boldsymbol{A} - \boldsymbol{B}\boldsymbol{K}_{\mathrm{P}} & -\boldsymbol{B}\boldsymbol{K}_{\mathrm{I}} \\ \boldsymbol{O} & \boldsymbol{O} & \boldsymbol{C} & \boldsymbol{O} \end{pmatrix} \begin{pmatrix} \boldsymbol{x}_{\mathrm{CT}}(t) \\ \boldsymbol{x}_{\mathrm{CTr}}(t) \\ \boldsymbol{x}_{\mathrm{s}}(t) \\ \boldsymbol{x}_{\mathrm{sr}}(t) \end{pmatrix} \\ &\quad + \begin{pmatrix} \boldsymbol{O} \\ -\boldsymbol{I}_r \\ \boldsymbol{O} \\ -\boldsymbol{I}_r \end{pmatrix} \boldsymbol{w}(t) + \begin{pmatrix} \boldsymbol{E} & \boldsymbol{O} \\ \boldsymbol{O} & \boldsymbol{O} \\ \boldsymbol{O} & \boldsymbol{E} \\ \boldsymbol{O} & \boldsymbol{O} \end{pmatrix} \begin{pmatrix} \boldsymbol{d}(t) \\ \hat{\boldsymbol{d}}_k \end{pmatrix}, \quad \begin{pmatrix} \boldsymbol{x}_{\mathrm{CT}}(t_k) \\ \boldsymbol{x}_{\mathrm{CTr}}(t_k) \\ \boldsymbol{x}_{\mathrm{s}}(t_k^+) \\ \boldsymbol{x}_{\mathrm{sr}}(t_k) \end{pmatrix} = \begin{pmatrix} \boldsymbol{x}_{\mathrm{CT}}(t_k) \\ \boldsymbol{x}_{\mathrm{CTr}}(t_k) \\ \boldsymbol{x}(t_k) \\ \boldsymbol{x}_{\mathrm{sr}}(t_k) \end{pmatrix}. \end{aligned}$$

With the state transformation

$$\begin{pmatrix} \boldsymbol{x}_{\mathrm{CT}}(t) \\ \boldsymbol{x}_{\mathrm{CTr}}(t) \\ \boldsymbol{x}_{\mathrm{s},\Delta}(t) \\ \boldsymbol{x}_{\mathrm{sr},\Delta}(t) \end{pmatrix} = \begin{pmatrix} \boldsymbol{I}_n & \boldsymbol{O} & \boldsymbol{O} & \boldsymbol{O} \\ \boldsymbol{O} & \boldsymbol{I}_r & \boldsymbol{O} & \boldsymbol{O} \\ \boldsymbol{I}_n & \boldsymbol{O} & -\boldsymbol{I}_n & \boldsymbol{O} \\ \boldsymbol{O} & \boldsymbol{I}_r & \boldsymbol{O} & -\boldsymbol{I}_r \end{pmatrix} \begin{pmatrix} \boldsymbol{x}_{\mathrm{CT}}(t) \\ \boldsymbol{x}_{\mathrm{CTr}}(t) \\ \boldsymbol{x}_{\mathrm{s}}(t) \\ \boldsymbol{x}_{\mathrm{sr}}(t) \end{pmatrix}$$

the model

$$\begin{aligned} \begin{pmatrix} \dot{\boldsymbol{x}}_{\mathrm{CT}}(t) \\ \dot{\boldsymbol{x}}_{\mathrm{CTr}}(t) \\ \dot{\boldsymbol{x}}_{\mathrm{s},\Delta}(t) \\ \dot{\boldsymbol{x}}_{\mathrm{sr},\Delta}(t) \end{pmatrix} &= \left(\begin{array}{cc|cc} \boldsymbol{A} - \boldsymbol{B}\boldsymbol{K}_{\mathrm{P}} & -\boldsymbol{B}\boldsymbol{K}_{\mathrm{I}} & \boldsymbol{O} & \boldsymbol{O} \\ \boldsymbol{C} & \boldsymbol{O} & \boldsymbol{O} & \boldsymbol{O} \\ \hline \boldsymbol{O} & \boldsymbol{O} & \boldsymbol{A} - \boldsymbol{B}\boldsymbol{K}_{\mathrm{P}} & -\boldsymbol{B}\boldsymbol{K}_{\mathrm{I}} \\ \boldsymbol{O} & \boldsymbol{O} & \boldsymbol{C} & \boldsymbol{O} \end{array}\right) \begin{pmatrix} \boldsymbol{x}_{\mathrm{CT}}(t) \\ \boldsymbol{x}_{\mathrm{CTr}}(t) \\ \boldsymbol{x}_{\mathrm{s},\Delta}(t) \\ \boldsymbol{x}_{\mathrm{sr},\Delta}(t) \end{pmatrix} \\ &\quad + \begin{pmatrix} \boldsymbol{O} \\ -\boldsymbol{I}_r \\ \boldsymbol{O} \\ \boldsymbol{O} \end{pmatrix} \boldsymbol{w}(t) + \begin{pmatrix} \boldsymbol{E} & \boldsymbol{O} \\ \boldsymbol{O} & \boldsymbol{O} \\ \boldsymbol{E} & -\boldsymbol{E} \\ \boldsymbol{O} & \boldsymbol{O} \end{pmatrix} \begin{pmatrix} \boldsymbol{d}(t) \\ \hat{\boldsymbol{d}}_k \end{pmatrix} \\ \begin{pmatrix} \boldsymbol{x}_{\mathrm{CT}}(t_k) \\ \boldsymbol{x}_{\mathrm{CTr}}(t_k) \\ \boldsymbol{x}_{\mathrm{s},\Delta}(t_k^+) \\ \boldsymbol{x}_{\mathrm{sr},\Delta}(t_k) \end{pmatrix} &= \begin{pmatrix} \boldsymbol{x}_{\mathrm{CT}}(t_k) \\ \boldsymbol{x}_{\mathrm{CTr}}(t_k) \\ \boldsymbol{x}_{\mathrm{CT}}(t_k) - \boldsymbol{x}(t_k) \\ \boldsymbol{x}_{\mathrm{CTr}}(t_k) - \boldsymbol{x}_{\mathrm{sr}}(t_k) \end{pmatrix} \end{aligned}$$

is obtained. The lower part of this model describes the difference behaviour

$$\begin{aligned} \begin{pmatrix} \dot{\boldsymbol{x}}_{\mathrm{s},\Delta}(t) \\ \dot{\boldsymbol{x}}_{\mathrm{sr},\Delta}(t) \end{pmatrix} &= \underbrace{\begin{pmatrix} \boldsymbol{A} - \boldsymbol{B}\boldsymbol{K}_{\mathrm{P}} & -\boldsymbol{B}\boldsymbol{K}_{\mathrm{I}} \\ \boldsymbol{C} & \boldsymbol{O} \end{pmatrix}}_{\bar{\boldsymbol{A}}_{\mathrm{I}}} \underbrace{\begin{pmatrix} \boldsymbol{x}_{\mathrm{s},\Delta}(t) \\ \boldsymbol{x}_{\mathrm{sr},\Delta}(t) \end{pmatrix}}_{\boldsymbol{x}_{\Delta\mathrm{I}}(t)} + \begin{pmatrix} \boldsymbol{E} & -\boldsymbol{E} \\ \boldsymbol{O} & \boldsymbol{O} \end{pmatrix} \begin{pmatrix} \boldsymbol{d}(t) \\ \hat{\boldsymbol{d}}_k \end{pmatrix} \\ \begin{pmatrix} \boldsymbol{x}_{\mathrm{s},\Delta}(t_k^+) \\ \boldsymbol{x}_{\mathrm{sr},\Delta}(t_k) \end{pmatrix} &= \begin{pmatrix} \boldsymbol{x}_{\mathrm{CT}}(t_k) - \boldsymbol{x}(t_k) \\ \boldsymbol{x}_{\mathrm{CTr}}(t_k) - \boldsymbol{x}_{\mathrm{sr}}(t_k) \end{pmatrix}, \end{aligned}$$

where $\bar{\boldsymbol{A}}_{\mathrm{I}}$ is Hurwitz and $\boldsymbol{d}(t) - \hat{\boldsymbol{d}}_k$ is bounded by assumption. Hence, the state difference $\boldsymbol{x}_{\mathrm{sr},\Delta}(t) = \boldsymbol{x}_{\mathrm{CTr}}(t) - \boldsymbol{x}_{\mathrm{sr}}(t)$ between the integrator states of the continuous-time PI-control loop and the event-based PI-control loop is bounded and a stationary bound can be derived as follows

$$\begin{aligned}
\lim_{t\to\infty} \|\boldsymbol{x}_{\Delta\mathrm{I}}(t)\| &= \left\| \int_{t_k}^{\infty} \mathrm{e}^{\bar{\boldsymbol{A}}_\mathrm{I}(t-\alpha)} \begin{pmatrix} \boldsymbol{E} \\ \boldsymbol{O} \end{pmatrix} (\boldsymbol{d}(\alpha) - \hat{\boldsymbol{d}}_k)\, \mathrm{d}\alpha \right\| \\
&\leq \int_0^{\infty} \left\| \mathrm{e}^{\bar{\boldsymbol{A}}_\mathrm{I}\alpha} \right\| \mathrm{d}\alpha \cdot \|\boldsymbol{E}\|\, \bar{d}_\mathrm{UD} \\
&= \bar{e} \cdot \int_0^{\infty} \left\| \mathrm{e}^{\bar{\boldsymbol{A}}_\mathrm{I}\alpha} \right\| \mathrm{d}\alpha \cdot \frac{\|\boldsymbol{E}\|}{\int_0^{\infty} \left\| \mathrm{e}^{\boldsymbol{A}\alpha} \right\| \mathrm{d}\alpha \cdot \|\boldsymbol{E}\|} \\
&= \bar{e} \cdot x_{\mathrm{srmax},\Delta},
\end{aligned}$$

where $\bar{d}_\mathrm{UD}$ was replace by Eq. (4.28). Consequently, the relation

$$\exists \bar{t} \geq t_k \text{ s.t. } \|\boldsymbol{x}_\mathrm{CTr}(t) - \boldsymbol{x}_\mathrm{sr}(t)\| \leq \bar{e} \cdot x_{\mathrm{srmax},\Delta} \text{ for } t \geq \bar{t}$$

holds, which together with Eq. (B.11) yields the bound $e_{\mathrm{max,PI}}$ defined in Eq. (4.32). □

B.9. Proof of Lemma 11

The lemma follows directly by considering event condition (4.40), (4.41) and Eq. (4.48). An event is generated if the difference state $\|\boldsymbol{x}_\Delta(\ell)\|$ exceeds the event threshold $\bar{e}$. Here, condition (4.41) is generally satisfied with equality sign at a continuous-time instance $\tilde{t}$ between the two consecutive sampling times $\ell_k - 1$ and ℓ_k (see Fig. 4.10, page 92). The maximum evolution of $\boldsymbol{x}_\Delta(t)$ in the remaining time interval $t \in [\tilde{t}, \ell_k T_\mathrm{s})$ can be overapproximated by an upper bound for the evolution of $\boldsymbol{x}_\Delta(\ell)$ in a single discrete-time step ($[(\ell_k - 1)T_\mathrm{s}, \ell_k T_\mathrm{s})$, Eq. (4.46))

$$\begin{aligned}
\|\boldsymbol{x}_\Delta(\ell_k) - \boldsymbol{x}_\Delta(\ell_k - 1)\| &= \|\boldsymbol{A}_\mathrm{DT}^{\ell_k-(\ell_k-1)} \boldsymbol{x}_\Delta(\ell_k - 1) \\
&\quad + \boldsymbol{E}_\mathrm{DT}(\boldsymbol{d}(\ell_k - 1) - \hat{\boldsymbol{d}}_{k-1}) - \boldsymbol{x}_\Delta(\ell_k - 1)\| \\
&\leq \|\boldsymbol{A}_\mathrm{DT} - \boldsymbol{I}_n\| \cdot \bar{e} + \|\boldsymbol{E}_\mathrm{DT}\| \gamma d_\mathrm{max} = x_{\mathrm{max,DTEB}},
\end{aligned}$$

which together with the event threshold $\bar{e}$ gives the right-hand side of Eq. (4.53). The term $\|\boldsymbol{x}_\Delta(\ell_k - 1)\|$ was replaced by the event threshold $\bar{e}$.

As at event times ℓ_k the update mechanism resets $\boldsymbol{x}_\mathrm{s}(\ell_k)$ according to $\boldsymbol{x}_\mathrm{s}(\ell_k^+) = \boldsymbol{x}(\ell_k)$ ($\boldsymbol{x}_\Delta(\ell_k^+) = \boldsymbol{0}$), the difference $\boldsymbol{x}(\ell) - \boldsymbol{x}_\mathrm{s}(\ell)$ remains in the set (4.53) for all sampling times $\ell \geq 0$. □

B.10. Proof of Theorem 13

For the state difference $\boldsymbol{e}(\ell) = \boldsymbol{x}(\ell) - \boldsymbol{x}_{\mathrm{DT}}(\ell)$, Eqs. (4.38), (4.42) and (4.44) yield

$$\begin{aligned} \boldsymbol{e}(\ell+1) &= \bar{\boldsymbol{A}}_{\mathrm{DT}}\boldsymbol{e}(\ell) + \boldsymbol{B}\boldsymbol{K}_{\mathrm{DT}}(\boldsymbol{x}(\ell) - \boldsymbol{x}_{\mathrm{s}}(\ell)), \\ \boldsymbol{e}(0) &= \boldsymbol{0} \end{aligned}$$

(cf. Section 3.4.1). As the difference $\boldsymbol{x}(\ell) - \boldsymbol{x}_{\mathrm{s}}(\ell)$ is bounded according to Lemma 11 and the discrete-time state-feedback loop is assumed to be stable, the state difference $\boldsymbol{e}(\ell)$ is bounded as well. Due to relation (4.53) and under the assumption $\|\bar{\boldsymbol{A}}_{\mathrm{DT}}\| < 1$, the inequalities

$$\begin{aligned} \|\boldsymbol{e}(\ell)\| &\leq \left\| \sum_{j=0}^{\ell-1} \bar{\boldsymbol{A}}_{\mathrm{DT}}^{\ell-1-j}\boldsymbol{B}\boldsymbol{K}_{\mathrm{DT}}(\boldsymbol{x}(j) - \boldsymbol{x}_{\mathrm{s}}(j)) \right\| \\ &\leq \sum_{j=0}^{\ell-1} \left\| \bar{\boldsymbol{A}}_{\mathrm{DT}}^{\ell-1-j}\boldsymbol{B}\boldsymbol{K}_{\mathrm{DT}} \right\| \cdot \|(\boldsymbol{x}(j) - \boldsymbol{x}_{\mathrm{s}}(j))\| \\ &\leq \sum_{j=0}^{\infty} \left\| \bar{\boldsymbol{A}}_{\mathrm{DT}} \right\|^{j} \|\boldsymbol{B}\boldsymbol{K}_{\mathrm{DT}}\| \cdot (\bar{e} + x_{\mathrm{max,DTEB}}) \\ &\leq \frac{1}{1 - \left\| \bar{\boldsymbol{A}}_{\mathrm{DT}} \right\|} \|\boldsymbol{B}\boldsymbol{K}_{\mathrm{DT}}\| \cdot (\bar{e} + x_{\mathrm{max,DTEB}}) = e_{\mathrm{max,DTEB}} \end{aligned}$$

hold and prove the theorem. □

B.11. Proof of Lemma 15

In the relevant time interval $[t_k, t_k + \tau_k)$, the relations

$$\begin{pmatrix} \dot{\boldsymbol{x}}(t) \\ \dot{\boldsymbol{x}}_{\mathrm{s}}(t) \end{pmatrix} = \begin{pmatrix} \boldsymbol{A} & -\boldsymbol{B}\boldsymbol{K} \\ \boldsymbol{O} & \bar{\boldsymbol{A}} \end{pmatrix} \begin{pmatrix} \boldsymbol{x}(t) \\ \boldsymbol{x}_{\mathrm{s}}(t) \end{pmatrix} + \begin{pmatrix} \boldsymbol{E} \\ \boldsymbol{O} \end{pmatrix} \boldsymbol{d}(t) + \begin{pmatrix} \boldsymbol{O} \\ \boldsymbol{E} \end{pmatrix} \hat{\boldsymbol{d}}_{k-1}$$

$$\begin{pmatrix} \dot{\boldsymbol{x}}_{\Delta}(t) \\ \dot{\boldsymbol{x}}_{\mathrm{s}}(t) \end{pmatrix} = \begin{pmatrix} \boldsymbol{A} & \boldsymbol{O} \\ \boldsymbol{O} & \bar{\boldsymbol{A}} \end{pmatrix} \begin{pmatrix} \boldsymbol{x}_{\Delta}(t) \\ \boldsymbol{x}_{\mathrm{s}}(t) \end{pmatrix} + \begin{pmatrix} \boldsymbol{E} \\ \boldsymbol{O} \end{pmatrix} \boldsymbol{d}_{\Delta}(t) + \begin{pmatrix} \boldsymbol{O} \\ \boldsymbol{E} \end{pmatrix} \hat{\boldsymbol{d}}_{k-1}$$

hold with the initial state

$$\begin{pmatrix} \boldsymbol{x}_{\Delta}(t_k) \\ \boldsymbol{x}_{\mathrm{s}}(t_k) \end{pmatrix} = \begin{pmatrix} \boldsymbol{x}(t_k) - \boldsymbol{x}_{\mathrm{s}}(t_k) \\ \boldsymbol{x}_{\mathrm{s}}(t_k) \end{pmatrix}.$$

The second model yields

$$\begin{aligned}\|\boldsymbol{x}_\Delta(t)\| &= \left\| \mathrm{e}^{\boldsymbol{A}(t-t_k)}(\boldsymbol{x}(t_k)-\boldsymbol{x}_\mathrm{s}(t_k)) + \int_{t_k}^{t} \mathrm{e}^{\boldsymbol{A}(t-\alpha)}\boldsymbol{E}\boldsymbol{d}_\Delta(\alpha)\,\mathrm{d}\alpha \right\| \qquad (B.12)\\ &\leq \left\|\mathrm{e}^{\boldsymbol{A}(t-t_k)}\right\| \cdot \|\boldsymbol{x}(t_k)-\boldsymbol{x}_\mathrm{s}(t_k)\| + \int_{t_k}^{t} \left\|\mathrm{e}^{\boldsymbol{A}(t-\alpha)}\boldsymbol{E}\right\| \mathrm{d}\alpha \cdot \gamma\, d_\mathrm{max}\\ &\leq \max_{\tau\in[0,\,\bar\tau)} \left\|\mathrm{e}^{\boldsymbol{A}\tau}\right\| \cdot \bar e + \int_0^{\bar\tau} \left\|\mathrm{e}^{\boldsymbol{A}\alpha}\boldsymbol{E}\right\| \mathrm{d}\alpha \cdot \gamma\, d_\mathrm{max}. \qquad (B.13)\end{aligned}$$

Next, the conditions (5.35), (5.36) are derived. No event is generated if

$$\|\boldsymbol{x}(t)-\boldsymbol{x}_\mathrm{e}(t)\| < \bar e \qquad (B.14)$$

holds. With

$$\begin{aligned}\boldsymbol{x}_\mathrm{e}(t) &= \mathrm{e}^{\bar{\boldsymbol{A}}(t-t_k)}\boldsymbol{x}(t_k) + \int_{t_k}^{t} \mathrm{e}^{\bar{\boldsymbol{A}}(t-\alpha)}\boldsymbol{E}\,\hat{\boldsymbol{d}}_k\,\mathrm{d}\alpha\\ \boldsymbol{x}_\mathrm{s}(t) &= \mathrm{e}^{\bar{\boldsymbol{A}}(t-t_k)}\boldsymbol{x}_\mathrm{s}(t_k) + \int_{t_k}^{t} \mathrm{e}^{\bar{\boldsymbol{A}}(t-\alpha)}\boldsymbol{E}\,\hat{\boldsymbol{d}}_{k-1}\,\mathrm{d}\alpha\end{aligned}$$

the state $\boldsymbol{x}_\mathrm{e}(t)$ can be replaced by

$$\boldsymbol{x}_\mathrm{e}(t) = \boldsymbol{x}_\mathrm{s}(t) + \mathrm{e}^{\bar{\boldsymbol{A}}(t-t_k)}(\boldsymbol{x}(t_k)-\boldsymbol{x}_\mathrm{s}(t_k)) + \int_{t_k}^{t} \mathrm{e}^{\bar{\boldsymbol{A}}(t-\alpha)}\boldsymbol{E}\,(\hat{\boldsymbol{d}}_k-\hat{\boldsymbol{d}}_{k-1})\,\mathrm{d}\alpha,$$

which yields the inequality

$$\left\|\boldsymbol{x}_\Delta(t) - \mathrm{e}^{\bar{\boldsymbol{A}}(t-t_k)}(\boldsymbol{x}(t_k)-\boldsymbol{x}_\mathrm{s}(t_k)) - \int_{t_k}^{t} \mathrm{e}^{\bar{\boldsymbol{A}}(t-\alpha)}\boldsymbol{E}\,(\hat{\boldsymbol{d}}_k-\hat{\boldsymbol{d}}_{k-1})\,\mathrm{d}\alpha\right\| < \bar e.$$

Hence, in order to guarantee $\tau_k < t_{k+1} - t_k$ for all k, the inequality

$$\begin{aligned}&\left\|\left(\mathrm{e}^{\boldsymbol{A}(t-t_k)} - \mathrm{e}^{\bar{\boldsymbol{A}}(t-t_k)}\right)\cdot(\boldsymbol{x}(t_k)-\boldsymbol{x}_\mathrm{s}(t_k))\right.\\ +&\left.\int_{t_k}^{t} \mathrm{e}^{\boldsymbol{A}(t-\alpha)}\boldsymbol{E}\boldsymbol{d}_\Delta(\alpha)\,\mathrm{d}\alpha - \int_{t_k}^{t} \mathrm{e}^{\bar{\boldsymbol{A}}(t-\alpha)}\boldsymbol{E}\,(\hat{\boldsymbol{d}}_k-\hat{\boldsymbol{d}}_{k-1})\,\mathrm{d}\alpha\right\| < \bar e\end{aligned}$$

must be satisfied, where $\boldsymbol{x}_\Delta(t)$ was replaced according to Eq. (B.12).

Due to assumption (5.34), the disturbance $\|\boldsymbol{d}_\Delta(t)\| = \|\boldsymbol{d}(t) - \hat{\boldsymbol{d}}_{k-1}\|$ and, thus, the difference $\|\hat{\boldsymbol{d}}_k - \hat{\boldsymbol{d}}_{k-1}\|$ are bounded by γd_max, but the differences inside the norms might have an opposite sign. Consequently, the delay bound $\tau^{\star d}$ can be determined by

$$\tau^{\star d} = \min \tau \geq 0 \text{ s.t. } \bar{a}(\tau) \cdot \bar{e} + d_{\text{xdd}}(\tau) \cdot \gamma\, d_{\max} = \bar{e}$$

with

$$\begin{aligned} \bar{a}(\tau) &= \left\| \mathrm{e}^{\boldsymbol{A}\tau} - \mathrm{e}^{\bar{\boldsymbol{A}}\tau} \right\| \\ d_{\text{xdd}}(\tau) &= \int_0^\tau \left(\left\| \mathrm{e}^{\boldsymbol{A}\alpha} \right\| + \left\| \mathrm{e}^{\bar{\boldsymbol{A}}\alpha} \right\| \right) \mathrm{d}\alpha \|\boldsymbol{E}\|. \end{aligned}$$

This leads to the conditions

$$\begin{aligned} \tau^{\star d} &= \arg \min_{\tau \in [0, \tilde{\tau}]} \left\{ \frac{\int_0^\tau \left(\left\| \mathrm{e}^{\boldsymbol{A}\alpha} \right\| + \left\| \mathrm{e}^{\bar{\boldsymbol{A}}\alpha} \right\| \right) \mathrm{d}\alpha \cdot \|\boldsymbol{E}\| \gamma\, d_{\max}}{1 - \left\| \mathrm{e}^{\boldsymbol{A}\tau} - \mathrm{e}^{\bar{\boldsymbol{A}}\tau} \right\|} = \bar{e} \right\} \\ \tilde{\tau} &= \arg \min_{\tau \geq 0} \left\{ \left\| \mathrm{e}^{\boldsymbol{A}\tau} - \mathrm{e}^{\bar{\boldsymbol{A}}\tau} \right\| = 1 \right\}, \end{aligned}$$

which together with relation (B.13) prove the lemma. □

C. List of symbols

General conventions

- **Scalars** are represented by lower-case italic letters (x, u, y).
- **Vectors** are represented by lower-case bold italic letters $(\boldsymbol{x}, \boldsymbol{u}, \boldsymbol{y})$.
- **Matrices** are represented by upper-case bold italic letters $(\boldsymbol{A}, \boldsymbol{B}, \boldsymbol{C})$.

Indices

$(.)_{\max}$	Maximum value of a scalar
$(.)_{\min}$	Minimum value of a scalar
$(.)^{-1}$	Inverse of a matrix
$(.)^{+}$	Pseudoinverse of a matrix
$(.)'$	Transpose of a vector or matrix
$\bar{(.)}$	Constant signal
$\dot{(.)}$	Time derivative of a signal
$(.)_0$	Initial value of a signal at time $t = 0$
$(.)_{\mathrm{CT}}$	Signal of the continuous-time control loop
$(.)_{\mathrm{DT}}$	Signal of the discrete-time control loop
$(.)_{\mathrm{s}}$	Signal of the control input generator
$(.)_{\mathrm{e}}$	Signal of the event generator, if $(.)_{\mathrm{s}} \neq (.)_{\mathrm{e}}$
$(.)_{\Delta}$	Difference signal
$\hat{(.)}$	Estimated signal

Scalars

n	Dimension of the state vector

m	Dimension of the input vector
l	Dimension of the disturbance vector
r	Dimension of the output vector
$\bar{e}$	Event threshold
t	Time
t_k	Event time
k	Event counter
t_ℓ	Discrete-time instance
ℓ	Discrete-time counter
τ	Delay
T_s	Sampling period
γ	Disturbance parameter
ξ	Quantisation level
q	Quantised signal

Vectors

$\boldsymbol{x}$	State vector
$\boldsymbol{u}$	Input vector
$\boldsymbol{y}$	Output vector
$\boldsymbol{d}$	Disturbance vector
$\hat{\boldsymbol{d}}$	Disturbance estimate
$\boldsymbol{d}_\Delta$	Disturbance estimation error
$\boldsymbol{d}_a$	Augmented disturbance
$\boldsymbol{w}$	Setpoint
$\boldsymbol{v}$	Measurement noise
$\boldsymbol{x}_s$	State of the control input generator
$\boldsymbol{x}_e$	State of the event generator, if $\boldsymbol{x}_s \neq \boldsymbol{x}_e$
$\boldsymbol{x}_\Delta$	Difference state ($\boldsymbol{x}_\Delta = \boldsymbol{x} - \boldsymbol{x}_s$)
$\hat{\boldsymbol{x}}$	Estimated state
$\boldsymbol{x}_{CT}$	State of the continuous-time control loop
$\boldsymbol{x}_{DT}$	State of the discrete-time control loop
$\boldsymbol{e}$	Approximation error ($\boldsymbol{e} = \boldsymbol{x} - \boldsymbol{x}_{CT}$)
$\boldsymbol{0}$	Null vector of appropriate dimension

Matrices

$\boldsymbol{A}$	System matrix
$\boldsymbol{B}$	Input matrix
$\boldsymbol{C}$	Output matrix
$\boldsymbol{E}$	Disturbance input matrix
$\boldsymbol{K}$	Controller matrix
$\boldsymbol{V}$	Prefilter
$\bar{\boldsymbol{A}}$	System matrix of a closed-loop system ($\bar{\boldsymbol{A}} = \boldsymbol{A} - \boldsymbol{BK}$)
$\boldsymbol{I}_n$	Identity matrix of size n
$\boldsymbol{O}$	Null matrix of appropriate size

Sets

$\mathbb{N}$	Set of nonnegative integers
$\mathbb{R}$	Set of real numbers
$\mathbb{R}_+$	Set of nonnegative real numbers
$\mathbb{R}^n$	Set of real vectors with dimension n
$\mathbb{R}^{n \times m}$	Set of real matrices with n rows and m columns
Ω_i	Stationary subset of the state space

Abbreviations

NCS	Networked control systems
GUUB	Globally uniformly ultimately bounded
ZOH	Zero-order hold
s.t.	Such that
SISO	Single input and single output

Persönliche Daten:

Name	Daniel Lehmann
Geburtsdatum	15.06.1981
Geburtsort	Essen

Bildungs- und Berufsweg:

Seit 06 / 2007	Wissenschaftlicher Mitarbeiter am Lehrstuhl für Automatisierungstechnik und Prozessinformatik (Prof. Dr.-Ing. Jan Lunze), Ruhr-Universität Bochum.
10 / 2001 - 04 / 2007	Studium der Elektrotechnik an der Ruhr-Universität Bochum mit den Schwerpunkten „Regelungs- und Energietechnik", Titel der Diplomarbeit: „Steuerungsentwurf für eine Klasse hybrider Systeme anhand eines mittleren Modells".
05 / 2006 - 08 / 2006	Praktikum bei der Vorwerk Elektrowerke GmbH & Co. KG, Abteilung für physikalische Grundlagen - Motoren, Wuppertal.
08 / 1991 - 06 / 2000	Helmholtz-Gymnasium Essen.
1987 - 1991	Grundschule, Essen.